T0205815

Macroevolutionary Theory on Macroecological Patterns

In *Macroevolutionary Theory on Macroecological Patterns*, Peter Price
establishes a completely new vision of the central themes in ecology.
For the first time in book form, the study of distribution, abundance,
and population dynamics in animals is cast in an evolutionary
framework. The book argues that evolved characters of organisms
such as morphology, behavior, and life history influence strongly their
ecological relationships, including the way that populations fluctuate
through time and space. The central ideas in the book are supported
by data gathered from over 20 years of research, primarily into plant
and herbivore interactions, concentrating on insects. The huge
diversity of insect herbivores provides the immense comparative power
necessary for a strong evolutionary study of ecological principles.

The book is intended as essential reading for all researchers
and students of ecology, evolutionary biology, and behavior, and for
entomologists working in agriculture, horticulture, and forestry.

PETER W. PRICE is Regents' Professor Emeritus at Northern Arizona
University, Flagstaff, U.S.A. Over the past 40 years Professor Price has
contributed over 200 research articles, and book chapters to the
scientific literature and has been sole author or an editor of 11 books.
He has received the Founder's Memorial Award from the
Entomological Society of America and is an Honorary Fellow of the
Royal Entomological Society.

Macroevolutionary Theory on Macroecological Patterns

PETER W. PRICE
Northern Arizona University

CAMBRIDGE
UNIVERSITY PRESS

University Printing House, Cambridge CB2 8BS, United Kingdom

Cambridge University Press is part of the University of Cambridge.

It furthers the University's mission by disseminating knowledge in the pursuit of education, learning and research at the highest international levels of excellence.

www.cambridge.org
Information on this title: www.cambridge.org/9780521520379

© Cambridge University Press 2003

First published 2003

A catalogue record for this publication is available from the British Library

Library of Congress Cataloguing in Publication data

Price, Peter W.
Macroevolutionary theory on macroecological patterns / by Peter W. Price.
p.   cm.
Includes bibliographical references (p.).
ISBN 0 521 81712 9 (hb.) – ISBN 0 521 52037 1 (pb.)
1. Macroevolution.   2. Ecology.   I. Title.
QH371.5 .P75   2003
576.8 – dc21   2002071489

ISBN 978-0-521-81712-7 Hardback
ISBN 978-0-521-52037-9 Paperback

Contents

Preface

The field of distribution, abundance, and population dynamics
has never fully embraced evolutionary theory as a guiding light or a
central theme. The field has remained largely ecological in its focus, in
spite of many other areas of ecology becoming more integrated with
evolutionary thought. However, taking an evolutionary view enables a
synthesis of many biological aspects of organisms and their ecology.
An integration of behavior, ecology, and evolution is essential in a full
understanding of any kind of interaction between a species and its
environment. Every species is molded by past events, selective forces,
and the "baggage" of its lineage. Hence the ecology of a species is very
much a function of evolved traits involving behavior, physiology, and
life history.

It is all the more surprising that population dynamics has remained largely aloof from evolutionary thinking when we recognize
that distribution, abundance, and dynamics of organisms have played
a central role in the development of ecology. The field is fundamental
in solving human problems with pest species in agriculture, horticulture, forestry, and epidemiology, and the history of ecology is full of
rich debates among major professional ecologists. These debates would
have been enriched, and perhaps resolved, if evolutionary points of view
were given equal play.

More than 20 years ago our research group began working on an
uncommon sawfly, about which very little was known, and its relatives
were poorly known also. The distribution, abundance, and population
dynamics turned out to be simple, for experimental approaches were
easy to apply, with strong limits on population size set by plant quality. With a low carrying capacity for the population set by severe resource limitation, we began to ponder why many species could escape
such limitation to become serious pests that could defoliate forests

on occasion. We adopted a completely novel perspective on why pest species existed. We entered into a comparison between uncommon and outbreak species to understand why the differences were so great. And an evolutionary approach proved to be the most enlightening, with the broadest possible comparative insights. After many idiosyncratic studies of individual species in various parts of the globe and a few synthesis papers, the time is right for a book-length treatment laying out the rationale, the thesis, the evidence, and the generality of the approach. A macroevolutionary basis, encompassing phylogenetic comparisons, is advocated for understanding big, macroecological patterns in distribution, abundance, and population dynamics. Of necessity, a broad view of the literature is required: to understand current views and our departure from these views, to show how broadly our approach applies to the literature, and to encompass a diverse array of taxa including species in temperate and tropical environments.

The approach is integrative in relation to the taxa it applies to, in bringing evolution, ecology, behavior, and life history to play equally important roles in understanding organisms, and the comparative approach is valid for temperate and tropical latitudes, as well as for multitrophic-level studies. The approach is Darwinian, starting with empirical observations in natural settings, finding patterns in nature, and generating hypotheses to explain mechanistically why such patterns should be evident.

Therefore, this book is intended for researchers and students in evolutionary biology, ecology, behavior, botany, and entomology and for entomologists working in agriculture, horticulture, or forestry. An attempt is made at applying the principles to plants and vertebrate groups near the end of the book. The main focus is on plant and herbivore interactions, concentrating on insects. Insect herbivores provide an almost unending richness of form, behavior, ecology, and dynamics, but also immense comparative power because of their diversity. This diversity is the key to a strong evolutionary foray into the bastions of ecology: distribution, abundance, and population dynamics.

Peter W. Price
Flagstaff, Arizona
May, 2002

Acknowledgments

During the writing of this book I have drawn on my experiences ix
over the past 50 years, for in the "olden days" schools provided excellent education in the sciences, especially botany and zoology in my case.
Hence, my ideas and knowledge have been influenced by hundreds of
important people in my life which I acknowledge mentally, but will not
mention by name. Certainly, my parents were most influential in developing my love of nature and my fascination with all wild things. Also
included in the most informative and interesting of friends have been
the many graduate and undergraduate students in my research programs at the University of Illinois, Urbana–Champaign and at Northern
Arizona University. Without them my research productivity would have
been minor both in volume and in scope. They have enriched my life
immeasurably with their knowledge and their friendship.

The research discussed in this book has been supported gener-
ously by the University of Illinois, Northern Arizona University, and
20 years of continuous funding from the National Science Foundation.
I am also most appreciative of the many opportunities to travel provided by agencies in other countries, a Guggenheim Fellowship and
a Fulbright Senior Scholar Award. My friends in other countries have
provided invaluable opportunities to broaden my experiences and opportunities for research. Especially important among these are research
groups in Finland, Japan, and Brazil with which I have enjoyed long-
standing collaborative research programs. To Drs. Jorma Tahvanainen
and Heikki Roininen at the University of Joensuu, Finland, Dr. Takayuki
Ohgushi at the University of Hokkaido, Sapporo, and Kyoto University,
Japan, and Dr. Geraldo Wilson Fernandes at the Universidade Federal de
Minas Gerais, Belo Horizonte, Brazil, I acknowledge my deep gratitude
for their hospitality, generosity, collaborative enthusiasm, friendship,
and the many memorable experiences in the field together. Without

them the broad comparative approach in this book would not have been possible.

Once I had prepared a draft, several reviewers were generous with their time, energy, and expertise in contributing ideas that improved the book. While we did not agree invariably, their reviews were thought-provoking, insightful, and challenging. Timothy G. Carr at Cornell University, Timothy P. Craig at the University of Minnesota, Peter J. Cranston at the University of California, Davis, and an anonymous reviewer all provided excellent constructive criticism. Further expertise was provided by Louella J. Holter at the Bilby Research Center, Northern Arizona University, whose competence, accuracy, and alacrity in word processing has rendered my scribbled manuscript into formal text on this occasion and on many others over the past 15 years or so. I have appreciated her outstanding services over these many years. I was fortunate to meet Ward Cooper, Commissioning Editor in the Biological Sciences at Cambridge University Press at Iguassu Falls, Brazil in 2000, at the International Congress of Entomology. A most pleasurable interaction has resulted for which I am grateful.

To my wife, Maureen, and all the plants (and insects) in my greenhouse and garden, I offer my unending affection and gratitude for providing a peaceful, nurturing environment in which to write this book.

1

The general thesis

The argument developed in this book is that much of the evolved nature of a species or higher taxon has a direct causative influence on the central issues concerning the ecology of that taxon: distribution, abundance, and population dynamics. Therefore, the macroevolutionary basis of a taxon is essential for understanding the fundamentals of ecology. This approach has not been advocated or subscribed to in the literature, neither in classical ecological texts such as Allee *et al.* (1949), Andrewartha and Birch (1954), and Odum (1959), nor in current volumes (e.g. Colinvaux 1993; Begon *et al.* 1996; Ricklefs 1997; Stiling 1998; Ricklefs and Miller 2000). More specialized approaches to population ecology emphasize direct environmental conditions rather than the overarching involvement of macroevolution (e.g., Royama 1992; Brown 1995; Den Boer and Reddingius 1996; Rhodes *et al.* 1996; Hanski and Gilpin 1997).

The study of the distribution, abundance, and population dynamics of species has been a central focus for ecologists for at least a century, as emphasized by Andrewartha and Birch (1954). "Ecology is the scientific study of the interactions that determine the distribution and abundance of organisms" (Krebs 1994, p. 3). Driven by pragmatism, the need to understand populations was prompted by burgeoning human populations (e.g. Malthus 1798; Verhulst 1838; Pearl and Reed 1920), plagues of agricultural pests (e.g. Waloff 1946), defoliating forest insects (e.g. Bodenheimer 1930; Schwerdtfeger 1941), human diseases, and the vectors of etiological agents (e.g. Smith and Kilbourne 1893; Zinsser 1935; Manson-Bahr 1963). Therefore, a paradigm shift in the conceptual basis of such central issues in ecology should be of consequence for the majority of ecologists.

MACROEVOLUTION AND MACROECOLOGY

The terms macroevolution and macroecology are established in the literature. Macroevolution denotes evolution above the species level: the origin of new species, genera, families, etc., and the resulting phylogenetic relationships among taxa. The benefit of a macroevolutionary approach to ecology is that phylogenetic relationships provide the strongest and most extensive patterns to be found in nature. A comparative macroevolutionary approach provides a powerful and encompassing method for discovering and understanding ecological patterns. Macroecology was defined by Brown (1995, p. 10) as "a way of studying relationships between organisms and their environment that involves characterizing and explaining statistical patterns of abundance, distribution, and diversity." In their original discussion of macroecology, Brown and Maurer (1989, p. 1145) emphasized its involvement with the "analyses of statistical distributions of body mass, population density, and size and shape of geographic range." Lawton (1999, 2000) embraced the term macroecology and the statistical nature of its methodology. However, as Root (1996, p. 1311) noted in his insightful review of Brown's book, "only a few kinds of data, on traits that are relatively easy and straightforward to measure (e.g. body mass, length of appendages, geographic range), are available in sufficient quantity for analysis." Such constraints limit the development of this field.

In this book the term macroecology is extended to its logical limit, involving the study of broad patterns in ecology. This definition incorporates the topics covered by Brown and Maurer and becomes equivalent in scope to the term macroevolution.

THEORY AND HYPOTHESIS

Setting distribution, abundance, and population dynamics in a macroevolutionary and macroecological framework places these central themes in ecology on a far larger scale than in the past, affording a strongly comparative approach to the understanding of broad patterns in nature. I define scientific theory simply as the mechanistic explanation of broad patterns in nature. The patterns must be empirical and the explanations must be factual. This is the Darwinian concept of theory and the Darwinian approach to the development of theory. Only with this Darwinian view will scientific theory in ecology achieve its potential of accounting for broad patterns in nature. Thus my use of the term theory is in the narrow sense of truly mechanistic explanation of broad patterns in nature, as in the modern theory of evolution. A clear distinction is made between the term hypothesis, being an idea in need of more testing, and a theory, which is factually based and well tested, with the weight of evidence consistent with the main thesis. Darwin provided a profoundly insightful hypothesis on evolutionary mechanisms, but with factually flawed mechanistic explanations for the origins of variation in populations and the hereditary process. Nevertheless, the empirical observations that variation in populations persisted and that traits were passed down through generations were sufficient to render his hypothesis the basis for the modern theory of evolution. This book starts with an hypothesis, presents information and methodologies that test the hypothesis, and ends with an argument supporting acceptance of the hypothesis as theory.

Theory based on empirical patterns and explanations contrasts with much of so-called theoretical ecology which is largely devoted to hypothetical investigation. "As with all areas of evolutionary biology, theoretical development advances more quickly than does empirical evidence," wrote Johnson and Boerlijst (2002, p. 86). My view is that empirical pattern detection is primary. This then motivates the search for mechanisms, and if the pattern is broad its combination with a mechanistic explanation results in theory. Therefore, empirical studies direct the development of theory – a fully Darwinian view.

One of the major problems with ecology today is the existence of too much data and not enough theory, too many hypotheses and not enough testing, too many models and not enough verification. "Ecology is awash with all manner of untested (and often untestable) models, most claiming to be heuristic, many simple elaborations of earlier untested models. Entire journals are devoted to such work, and are as remote from biological reality as are faith-healers" (Simberloff 1980, p. 52). Models and hypothetical theory can be readily defended (e.g. Caswell 1988), but development of factually and empirically based broad patterns and their mechanistic understanding must surely advance the science of ecology more rapidly than any other component in this scientific endeavor.

Factual theory in ecology must cope with the tremendous diversity of organisms and phylogenetic pathways, recognizing that several to many outcomes are possible because of evolutionary and ecological processes. Theory must be pluralistic. Beginning with taxon A under ecological conditions B, the outcome will be C. With taxon A in different conditions D, the result may be E (cf. MacArthur 1972a). Outcomes are obviously conditional on the inputs and prevailing conditions, so that we should anticipate different results when different organisms evolve in the same environment or if the same organisms evolve in different environments. Theory must recognize the different phylogenies and conditions in which member species have evolved, and embrace pluralism as much as is needed, dictated by the relative conformity or diversification of the taxa under study. Ending their critique of the adaptationist or "Panglossian paradigm," Gould and Lewontin (1979, p. 597) endorsed the pluralistic approach:

> We welcome the richness that a pluralistic approach, so akin to Darwin's spirit, can provide. Under the adaptationist programme, the great historic themes of developmental morphology and *Bauplan* were largely abandoned; for if selection can break any correlation and optimize parts separately, then an organism's integration counts for little. Too often, the adaptationist programme gave us an evolutionary biology of parts and genes, but not of organisms. It assumed that all transitions could occur step by step and understated the importance of integrated developmental blocks and pervasive constraints of history and architecture. A pluralistic view could put organisms, with all their recalcitrant, yet intelligible, complexity, back into evolutionary theory.

This, in my view, is precisely what is needed in ecology. A pluralistic view, recognizing patterns resulting from different phylogenetic origins and *Baupläne*, and the macroevolutionary divergence of lineages, will bring ecology into a central place in evolutionary biology. Unless we embrace a macroevolutionary view of ecology we will remain collectors of facts, piles of facts, without theory to guide progress. We have piles of studies on plant and herbivore interactions, chemical ecology, and multitrophic-level interactions, but extraordinarily little pattern detection and certainly no factually based theory that is broadly supported and widely subscribed to: "a pile of sundry facts – some of them interesting or curious but making no meaningful picture as a whole" (Dobzhansky 1973, p. 129).

The field of ecological morphology is already well established. "Ecological morphology is broadly concerned with connections between how organisms are constructed and the ecological and evolutionary consequences of that design" (Reilly and Wainwright 1994, p. 339). The explicit assumption is that morphology has direct effects on ecology, a view heretofore absent in the sciences relating to population dynamics.

QUESTIONS

If we are to address broad patterns in nature and the underlying mechanisms driving pattern, there must be a set of broad questions to focus upon. These are far broader than generally conceived, especially relating to population dynamics. For example, why are some insect taxa replete with serious pest species, such as the short-horned grasshoppers, while others are full of innocuous and inconspicuous species, such as the tree hoppers? The acridid grasshoppers include the worst pests on earth in the form of plague locusts, but tree hoppers or membracids hardly enter into books on harmful insects. Even more closely related taxa can exemplify very different patterns of distribution, abundance, and population dynamics. We may well ask, why does one group so frequently show epidemic or outbreak dynamics, such as the pine sawflies, while its sister taxon contains very few outbreak species, as in the common sawflies? The pine sawfly family, Diprionidae, includes in North America almost 85 percent of species that are serious forest pests (Arnett 1993), but the family of common sawflies, Tenthredinidae, contains only about 3 percent that are regarded as pests (Price and Carr 2000).

Following such questions on broad patterns in nature there are the obvious additional questions on mechanisms. Why are outbreak, eruptive, or pest species so different in their population ecology from the many species that are patchily distributed, of low abundance over a landscape, and with relatively stable population dynamics? Why are some phylogenetically divergent taxa so similar in their population ecology? Specific taxa will be used to address and resolve these questions.

THE CENTRAL HYPOTHESIS

We have called our thesis the **Phylogenetic Constraints Hypothesis** (Price 1994b; Price *et al.* 1995a, 1998a; Price and Carr 2000). Its conceptual framework is developed best in Price and Carr (2000). The empirical observations and experiments, and the discovery of natural patterns, which initially prompted development of the hypothesis, are described in Chapters 3–5. The hypothesis argues that macroevolutionary patterns provide the mechanistic foundation for understanding broad ecological patterns in nature involving the distribution, abundance, and population dynamics of species and higher taxa. A **phylogenetic constraint** is a critical plesiomorphic character, or set of characters, common to a major taxon, that limits the major adaptive developments in a lineage and thus the ecological options for the taxon. However, many minor adaptations become coordinated to maximize the ecological potential of a species within the confines of the phylogenetic constraint. This set of adaptations constitutes the **adaptive syndrome** of the group. The adaptive syndrome has inevitable ecological consequences, named **emergent properties**, involving distribution, abundance, and population dynamics.

This hypothesis differs fundamentally from existing approaches to population ecology. Most current hypotheses are ecological and idiosyncratic, based on the study of single species, exemplifying the **idiographic program**, as expanded upon in the next chapter. The Phylogenetic Constraints Hypothesis is evolutionary, strongly comparative, and synthetic in its treatment of taxonomic groups higher than the species level, emphasizing basic mechanistic processes and broad patterns in nature: a truly **nomothetic approach** to population ecology: the **Macroevolutionary Nomothetic Paradigm**.

The terms we use in the Phylogenetic Constraints Hypothesis are established in the literature, although the mechanistic pathway of cause and effect is new, starting with our treatment in Price *et al.* (1990). McKitrick (1993, p. 309) defined a phylogenetic constraint as "any result or component in the phylogenetic history of a lineage that prevents an anticipated course of evolution in that lineage." Thus, a constraint limits the adaptive radiation of a lineage in a certain manner, such that the full potential radiation is not achieved. Such constraints are likely to have **phylogenetic effects** in the sense of Derrickson and Ricklefs (1988), meaning that closely related organisms are likely to be similar in their evolved characters of morphology, physiology, behavior, life history, and ecology.

The term adaptive syndrome was coined by Root and Chaplin (1976). "As organisms perfect a mode of life, their evolution is channeled so that a variety of adaptations are brought into harmony" (p. 139). This integrated set of adaptations was defined by Eckhardt (1979, p. 13) as "the coordinated set of characteristics associated with an adaptation or adaptations of overriding importance, e.g. the manner of resource utilization, predator defense, herbivore defense, etc." We use the term in this sense while arguing that the adaptive syndrome we assert to be central is that in relation to the phylogenetic constraint. That is, the syndrome is a set of adaptations that mitigate the effects of the constraint and may even turn it into some kind of advantage. As Ligon (1993, p. 3) said, "yesterday's adaptation may be today's constraint," but the reverse is also true.

An emergent property is one that arises as a natural or logical consequence or outcome. Brown (1995) used the term in this way. The term is often used, also, as a property that is unexpected and not predicted based solely on the knowledge of components, as in the combination of hydrogen and oxygen to make water (cf. Mayr 1982). For ecologists, Salt (1979, p. 145) recommended the operational definition: "An emergent property of an ecological unit is one which is wholly unpredictable from observation of the components of that unit." Whichever definition is preferred, our use complies. The argument we make that major patterns in distribution, abundance, and population dynamics are driven by mechanisms dictated by the evolved phylogenetic baggage of lineages is clearly unexpected based on the relevant literature discussed in Chapter 2. The argument can be developed only after detailed study of one species and its relatives, and must be based on a clear understanding of the evolutionary biology of the group. These points are covered at length in Chapters 3 and 4. And, just as we can now confidently predict that oxygen mixed with hydrogen will yield water, we can logically predict much of the ecology of a taxon based on its phylogenetic constraints and adaptive syndrome. In fact, our research program is akin to that of Sih *et al.* (1998) on emergent impacts of multiple predator effects (MPEs) on prey. They note that "Ultimately, our goal is not just to document the existence of emergent MPEs but to identify characteristics of predators, prey or the environment that tend to make one type of emergent effect... more likely than another" (Sih *et al.* 1998, p. 354). This is precisely the research focus of our program over the past decade, but relating to the emergent properties of population dynamics (e.g. Price *et al.* 1990, 1995a, 1998a; Price 1994b).

Ideally we should adopt a formal phylogenetic analysis of a clade mapping evolved traits on the phylogenetic hypothesis and the correlated emergent properties concerning distribution, abundance, and population dynamics. Such an analysis is not yet possible for any group because especially the population dynamics of many species in a taxon is not adequately documented. However, for the first time we do map population-level traits on a phylogeny, showing the causal linkage from a phylogenetic constraint to the adaptive syndrome to the emergent properties. Although the criteria used for the emergent properties are subjective more than quantitative, this first example provides a methodology for a rigorous test of the Phylogenetic Constraints Hypothesis (see Chapter 5).

Although the flow of effects from phylogenetic constraints to adaptive syndrome and its emergent properties forms the central theme of our thesis, at all steps resources intersect with evolutionary developments. Thus, the nature of resources utilized by insect herbivores must be understood in detail. Indeed, the display of resources may even override strong phylogenetic constraints, resulting in divergent adaptive syndromes and emergent properties, as explained in Chapter 8. However, I consider such strong effects to be unusual and more subtle influences to be more general among herbivorous insects. For example, differences in resources for the herbivorous Hymenoptera, sawflies and woodwasps, result in dramatic variation in ovipositor morphology but the *Bauplan*, including the lepismatid form of ovipositor, remains intact throughout. Among the most important biological features in the Hymenoptera, Gauld and Bolton (1988, p. 8) have the "ovipositional mechanism" first. But the interplay of the hymenopteran ovipositor and resource heterogeneity becomes a central issue in Chapter 3.

The novelty of the arguments developed in the Phylogenetic Constraints Hypothesis can be evaluated only in the light of past and current general views on the factors that influence the population dynamics of species. Therefore, I provide a brief historical overview of the field in Chapter 2. Then I progress to coverage of the focal species on which this hypothesis was developed, in Chapters 3 and 4, and to related species in Chapter 5. The importance of comparative studies across a taxon must be emphasized if we are to search for general patterns, mechanisms, and empirically based theory. The comparative approach is then extended in Chapter 6 to other taxa with similar constraints but more divergent phylogeny. Very different species with different dynamics are discussed in Chapter 7, sister taxa with divergent emergent properties are discussed in Chapter 8, and an attempt is made to advance the hypothesis into the world of vertebrates and plants in Chapter 9. Finally, Chapter 10 is devoted to a synthesis on the distribution, abundance, and dynamics of organisms.

2

Historical views on distribution, abundance, and population dynamics

A brief historical perspective on demography, distribution, abun-   9
dance, and population dynamics is essential for an appreciation of the
paradigmatic shift advocated in this book. Such a view is provided in
various sources by experts, which can be consulted for details, and from
which I have constructed some of the scenario presented here. In their
Principles of animal ecology, W. C. Allee, Alfred E. Emerson, Orlando Park,
Thomas Park, and Karl P. Schmidt (1949) devoted Section I to the history
of ecology up to 1942. Their authoritative view is valuable because they
had experienced first hand much of the development of ecology during
the twentieth century. Two books published in 1954 also became clas-
sics in ecology: David Lack's *The natural regulation of animal numbers* and
The distribution and abundance of animals by H. G. Andrewartha and L. C.
Birch, providing these authors' perspectives on the state of the field in
the mid 1950s. LaMont Cole (1957) wrote an excellent review on the
history of demography, and Tamarin (1978), in the Benchmark Papers
in Ecology series, provided a balanced treatment on *Population regulation*
with readings covering major points of view through the controversial
1960s and into the early 1970s. From the early 1960s I have worked in
this field of ecology, so I will provide a more personal view of developments since then. First, I will concentrate on how ideas developed into
the 1950s based on field studies and other empirical methods. Then
I will discuss demography and the emergence of life table analysis,
followed by trends up to the present day.

EARLY FIELD STUDIES

Because my prime concern in population dynamics is the insects, and
the study of insects has provided the basis for my macroevolutionary
approach, a gratifying detail is the early concentration of population

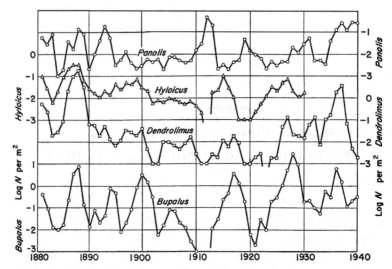

Fig. 2.1. Population densities of four forest insects in Letzlinger Heide,
Germany, sampled from 1880 to 1940. The species were the pine beauty
moth, *Panolis flammea* (Noctuidae), the pine hawk moth, *Hyloicus pinastri*
(Sphingidae), the pine spinner moth, *Dendrolimus pini* (Lasiocampidae),
and the bordered white moth, *Bupalus piniarius* (Geometridae). Densities,
plotted on a logarithmic scale, are winter census estimates of number of
moth pupae per m^2 in the soil for *Panolis*, *Hyloicus*, and *Bupalus*, and the
number of hibernating larvae per m^2 of forest floor for *Dendrolimus*.
(From Varley, G. C., G. R. Gradwell, and M. P. Hassell (1973) *Insect
population ecology*, Fig. 8.2, Blackwell Science, Oxford; based on Varley
1949.)

studies on forest Lepidoptera in Germany. Chronology of outbreaks was
recorded for major insect species starting in 1801 in Bavarian areas
with a record for 188 years (Klimetzek 1990). As a subset of these surveys, four species of moth were censused for 60 years from 1880 to
1940 (Fig. 2.1), although as in many studies since, a mechanistic understanding of the fluctuations was not achieved (Schwerdtfeger 1935,
1941). Schwerdtfeger rejected any simple explanation such as weather
or parasitic wasps and flies, recognizing that many factors may be important and each may affect the four species in different ways and
at different times. Varley (1949) attempted a new analysis of the data,
but concluded that insufficient data were provided for an informed
interpretation. "Let us hope that further work will be concentrated on
producing the detailed mortality and fertility data which may eventually help to provide a proper explanation of these fascinating problems"

(p. 122). No doubt, Varley took his own hopes to heart in designing his study, with Gradwell, on the winter moth, *Operophtera brumata*, which ran for 19 years from 1950 to 1968 (Varley *et al.* 1973; Hunter *et al.* 1997). Varley and Gradwell actually selected the trees in 1949 (Hassell *et al.* 1998), the year of publication of Varley's re-examination of the German forest lepidopterans.

Already, in these early studies and in Varley's inspection of the results, we observe some biases in the way scientists viewed empirical results. First, no mention was made by Schwerdtfeger or Varley of the possible role of variation in plant quality as a factor in moth population change. Second, while the problem of very low numbers between outbreaks was recognized, making their study practically impossible during these lows, no initiatives were developed to study alternative species with more persistent and higher general population levels. Third, a bias crept into Varley's views in which he expressed the opinion that "There is some indirect evidence that parasites themselves may be responsible for the gradations" (Varley 1949, p. 121). He justified this point by noting a very high percentage of parasitism accompanying the rapid declines in numbers. Fourth, the arguments are developed purely with an ecological perspective: those factors that might impinge directly on the reproduction and survival of individuals. No evolutionary scenarios were considered.

These kinds of thinking and approaches are what, I believe, shape the development of science and I am interested in following lines of reason into the present day as a way of capturing the development of ecology. We will see later in this chapter how Varley's own studies developed and how interpretations were considerably modified by Hunter. We will also note how the same biases mentioned above persisted into the recent past and even to the present day.

In their treatment of "economic biology," Allee *et al.* (1949) emphasized the importance of the study of insect pests of agricultural crops and trees, and efforts at biological control using the natural enemies of exotic pest species. For example, Howard (1897, p. 48) noted the way in which parasitic insects appeared to be regulating populations of the whitemarked tussock moth, *Orgyia leucostigma*. "With all very injurious lepidopterous larvae...we constantly see a great fluctuation in numbers, the parasite rapidly increasing immediately after the increase of the host species, overtaking it numerically, and reducing it to the bottom of another ascending period of development." Lotka (1924) actually cited this passage, being clearly impressed with "this oscillatory process" (p. 91), which he mimicked in his "predator–prey" equations. There is little doubt about the profound effect that Lotka, with Volterra, had on the view of top-down regulation of herbivores by carnivores. And the now legendary success of the biological control program on cottony-cushion scale, *Icerya purchasi*, in California, using the ladybird beetle or vedalia beetle, *Vedalia cardinalis*, also promoted the concept of top-down regulation of insect herbivore numbers (DeBach 1964, 1974). From its discovery in 1888 to complete control of all infestations in California in 1890, coupled with the rarity of the scale in its natural habitat in Australia, there was a strong indication that herbivorous insects could be relegated to rarity by natural enemies.

Studies of native pests on crops repeatedly suggested top-down regulation, reinforcing the view that biotic forces were of paramount importance in the regulation of animal numbers. An example involves the cotton boll weevil, *Anthonomous grandis*, studied by Pierce *et al.* (1912). They concluded that "The control of the boll weevil by insect enemies is sufficiently great to give it a high rank in the struggle against the pest. A considerable portion of the insect control would not be accomplished by any other factor" (p. 94). The complexities of biotic interactions, based on the cotton plant and involving the boll weevil, were emphasized in a figure remarkable for its time because of its broad community ecology and food web approaches (Fig. 2.2).

The school of thought that espoused population regulation of animals by natural enemies developed rapidly in the early decades of the 1900s. Tamarin (1978) selected two examples to illustrate this point: Howard and Fiske (1911) and Nicholson (1933). Howard and Fiske were involved with the biological control of gypsy moth and brown-tail moth, which were exotics in North America. Their interest in importing parasitic insects was founded on the principles that parasitic insects were the only ingredient of regulation that was missing compared to the Old World habitat, that insect parasites were specific to their host and thus dependent on the host as a source of food, and that parasites were therefore "facultative," or responsive in a density-dependent manner, to use Smith's (1935) terminology.

> Each species of insect in a country where the conditions are settled is subjected to a certain fixed average percentage of parasitism, which, in the vast majority of instances and in connection with numerous other controlling agencies, results in the maintenance of a perfect balance. The insect neither increases to such abundance as to be affected by disease or checked from further multiplication through lack of food, nor does it become extinct, but throughout maintains a degree of abundance in relation to other species existing in the same vicinity, which, when averaged

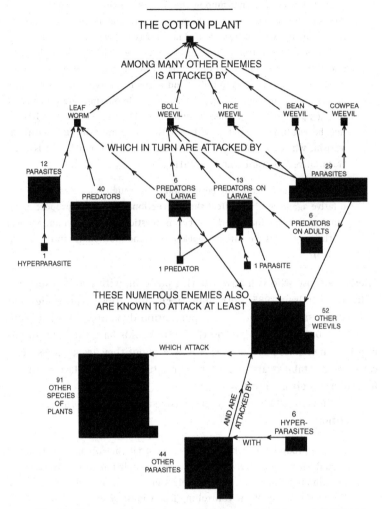

Fig. 2.2. The complex of interactions discovered by Pierce *et al.* (1912) for the boll weevil, *Anthonomous grandis*, and other herbivores on cotton, *Gossypium hirsutum*. This was regarded as a novel way to illustrate ecological interactions in the early 1900s by Allee *et al.* (1949), illustrating an early food web without the current convention of placing the plant at the base of the figure. The many parasitoids and predators recorded attacking the boll weevil caused Pierce *et al.* to conclude that biotic, top-down regulation of numbers was a crucial component of the system. (From Pierce *et al.* 1912.)

for a long series of years, is constant...In order that this balance may exist it is necessary that among the factors which work together in restricting the multiplication of the species there shall be at least one, if not more, which is what is here termed facultative (for want of a better name), and which, by exerting a restraining influence which is relatively more effective when other conditions favor undue increase, serves to prevent it. There are a very large number and a great variety of factors of more or less importance in effecting the control of defoliating caterpillars, and to attempt to catalogue them would be futile, but however closely they may be scrutinized very few will be found to fall into the class with parasitism, which in the majority of instances, though not in all, is truly "facultative." (Howard and Fiske 1911, p. 107)

A natural balance can only be maintained through the operation of facultative agencies which effect the destruction of a greater proportionate number of individuals as the insect in question increases in abundance. Of these agencies parasitism appears to be the most subtle in its action. (p. 108)

There are those who still advocate this position, with good reason (e.g. Munster-Swendsen 1985; Berryman 1999). But, after a career devoted to host–parasitoid interactions in population dynamics, we must note Hassell's (2000, p. 7) recognition that "parasitoids have the potential to regulate their host populations: whether or not they are a wide-spread cause of the stability and persistence of natural host populations in the field is still legitimately debated."

Nicholson (1933) chose to emphasize another biotic interaction – competition.

For the production of balance, it is essential that a controlling factor should act more severely against an average individual when the density of animals is high, and less severely when the density is low. In other words, the action of the controlling factor must be governed by the density of the population controlled...A moment's reflection will show that any factor having the necessary property for the control of populations must be some form of competition. (p. 135)

In his influential book on *The natural regulation of animal numbers*, Lack (1954) considered both reproductive rates and mortality in animals, covering birds, mammals, fish, and insects. He concluded that "Plant-eating insects are for most of the time held in check by insect parasites, but occasionally there is a violent increase which the parasites cannot check, and the numbers are then brought down by starvation" (p. 276). He thought the evidence indicated that food shortage was the most common limiting factor for many birds, mammals, and larger marine fish and some predatory insects. Predation was thought to be important for gallinaceous birds, but disease was implicated in red grouse populations and in humans. Both bottom-up factors of food supply and top-down factors of predators or parasites were invoked by Lack, but all were biotic in character.

Another intriguing facet of Lack's conclusions was his argument that mortality factors were more important than natality or reproductive rates in the regulation of animal numbers. "Summarizing, reproductive rates are a product of natural selection and are as efficient as possible. They may vary somewhat with population density, but the main density-dependent control of numbers probably comes through variation in the death-rate. The critical mortality factors are food shortage, predation, and disease, one of which may be paramount, though they often act together" (p. 276).

Thus, it is somewhat paradoxical that the one scientist who espoused evolutionary principles while considering population regulation, rejected the notion that overall they are important. Lack may well be considered as the founding father of evolutionary ecology, starting with his treatment of the clutch-size debate (Lack 1946, 1947a, b, 1948a, b, c; see also Collins 1986), but he certainly paved the way for the concentration exclusively on ecological factors in the study of population dynamics. Lack's view was no doubt prevalent, for in that year Morris and Miller (1954) independently wrote a most influential paper on life table construction. This approach emphasized the factors that depleted a cohort of animals through a generation from the egg stage to the adult. Thus, my preference is to call them "death tables" (Price 1997, p. 326). We will pick up the story of life tables or death tables later in this chapter.

As noted earlier in this chapter, a strong bias had developed by the 1950s among some scientists that biotic forces are paramount in population regulation: top-down effects of natural enemies, lateral effects of competition, and bottom-up impact involving a shortage of food. Conventional wisdom repeatedly emphasized mortality factors. Natality was rarely considered, except when Varley wished for information on this topic in order to understand Schwerdtfeger's data better and when Lack concluded that reproductive rates were efficient and of little consequence in the natural regulation of animals. Uvarov (1931, p. 174) had noted another general trend, stating that "entomologists of the present day are mainly historians of insect pests, in so far as they record and describe their outbreaks, and sometimes attempt to

analyse their causes." Emphasis was on description of individual pest
species outbreaks, not on a mechanistic understanding of eruptions
nor on a comparative approach among species. I would call this the
Idiosyncratic Descriptive Paradigm on insect population studies, or
the **Microecological Descriptive Paradigm**, of course driven by the pragmatic interest in economic agriculture and forestry. "For the practical
purpose of managing nature, ecology may be idiographic and not nomothetic" (Simberloff 1980, p. 51). However, the approach served the need
poorly because, without a truly mechanistic understanding of population dynamics and a broad comparative approach, the best methods
for management could not be determined, nor could they be applied
broadly. This Idiosyncratic Descriptive Paradigm has prevailed largely
to the present day.

Another aspect of developing views on distribution, abundance,
and population dynamics was, naturally enough, that each author was
influenced by the organisms studied and the main focus of their in-
quiry. Of course, those promoting biological control of insects or plants
would emphasize biotic top-down forces. When resources are clearly
limiting, such as for carrion flies, then competition would be recognized as critical in regulation, as in the case of Nicholson's (1954) studies on the sheep blowfly, *Lucilia cuprina*. When concern is focused on the
geographic distribution of organisms, the reason for the initial phases
of a population flush, or on eruptive species in dry climates such as
plague grasshoppers, then weather and climate become of prime interest. Emphasizing the disparities of interests, focal interests, and organisms largely diffuses the debates on the relative importance of biotic
and abiotic forces in population regulation. But such emphasis also promotes the search for pluralistic theory on the one hand, and concern
for careful adherence to a particular process when working compara-
tively. Pluralistic theory admits that there are several to many major
scenarios for any one central question, be it population regulation, geographic distribution, timing of outbreaks, or the relative importance
of bottom-up, lateral, or top-down effects in dynamics. When one group
emphasizes population regulation (exemplified by Nicholson and Bailey
1935; Nicholson 1954, 1957) and another group emphasizes geographic
distribution (exemplified by Andrewartha and Birch 1954; Birch 1958),
of course, unity is unlikely.

Just having alluded to abiotic factors such as weather and climate
opens up another major facet in early empirical studies that directed
researchers' attention. Tamarin (1978) liked to contrast the biotic and
abiotic schools of population regulation. But while there was some

friction among practitioners, as I stated above, I prefer to view the different emphases as deriving from different approaches, questions addressed, and organisms of prime interest. Bodenheimer (1925a, b, 1926, 1927) emphasized the role of weather in influencing developmental rates of insects, number of generations per year, and distributional boundaries of insects, all factors contributing to the distribution and abundance of insects. Such a view is hardly controversial. Uvarov (1931, 1966, 1977) was also very concerned with weather because of his interest in grasshoppers and especially plague species such as the desert locust, *Schistocerca gregaria*. Wind patterns were of concern because winds carried plagues with them, and rainfall patterns in time and space defined the local carrying capacity of forage for the grasshoppers in normally very dry climates. High rainfall in certain areas would provide enough green forage for three grasshopper generations over which time populations could pass from the solitary to the gregarious phase and enter into swarming, plague proportions (cf. Showler 1995). In a review on population dynamics of grasshoppers, Dempster (1963) concluded that top-down effects of natural enemies seldom exerted more than a small effect at high grasshopper densities. In a similar vein, the arguments about abiotic factors made by Andrewartha and Birch (1954) were equally valid and compelling, so long as we remember their focus.

Andrewartha and Birch (1954) addressed two questions: (1) Why does this animal inhabit so much and no more of the earth? (2) Why is it abundant in some parts of its distribution and rare in others? When geographical barriers are absent, we fully expect to observe more or less normal distributions of species over latitudinal, longitudinal, and altitudinal gradients (e.g. Brown 1995; Brown and Lomolino 1998; Price *et al.* 1998b). And the variables concerned in defining such distributions are usually precipitation, temperature, humidity, and their effects directly, or on the resources required for a species. Indeed, global vegetation types can be classified in the Holdridge system, employing only "mean annual biotemperature in degrees centigrade" (i.e. temperature above freezing) and "humidity provinces" based on the interaction of precipitation and temperature (Holdridge 1947, 1967; Holdridge *et al.* 1971). The pattern-generating effects of temperature work on both latitudinal and altitudinal gradients. The ecological niches of organisms are commonly viewed as distributions on one or more gradients, such as in Hutchinson's (1957) n-dimensional hypervolume and the Whittaker *et al.* (1973) combination of niche, habitat, and ecotope. Generally, one or more physical factors enter into the definition of the niche.

One can hardly doubt the veracity of the argument that the answers to Andrewartha and Birch's central questions must involve much about the physical factors of the environment. But, we can equally well cite many cases in which food supply or natural enemies or both are biotic forces of paramount importance in population dynamics. So behaving and writing as if there were two schools of thought, diametrically opposed and at daggers drawn, seems to create a red herring that should not have become a major debate in the 1950s and 1960s. Many authors have recognized the need for a balanced view admitting that both biotic and abiotic influences can be important in distribution, abundance, and population dynamics (e.g. Thompson 1929, 1939; Milne 1957a, b, 1962; Reynoldson 1957; Bakker 1964) while noting my own view that if you start with different questions and emphases the answers will also differ (e.g. Bakker 1964). Through all of this debate a surfeit of descriptive ecology prevailed while strongly mechanistic approaches using field experiments were undersubscribed (cf. Krebs 1995). As the reader will note, I have not raised the issue of the debates over density-dependent and density-independent factors in regulation or limitation of populations. The issues are well known and more than enough has been written on the subject. My view is that we should let empirical studies enlighten us on what really happens in nature in each particular case and in any efforts at generalization. The proposal by Berryman (1999) to use alternative language should be favored.

The points I wish to emphasize in this section on early field studies, covering trends into the 1950s, are as follows.

1. Plant quality variation in space and time was not considered as a critical factor in distribution, abundance, and population dynamics.

2. The behavior of organisms, especially females during reproduction, was not studied in most groups, the exception being birds in which territoriality and feeding of nestlings was obviously critical.

3. Little or no experimentation was used to examine the mechanisms driving population change and distribution.

4. Hypothesis erection and testing was not a mode of enquiry in most studies.

5. Very little synthesis was undertaken on patterns in distribution, abundance, and population dynamics.

6. An evolutionary approach was almost never taken, except for Lack's initiatives in evolutionary ecology.

The last four items in this list all contributed to what I have called the Idiosyncratic Descriptive Paradigm on population dynamics. Each population was studied for a particular reason, among fish and insects mostly for economic reasons, and among birds because of intrinsic interest. Lacking strong conceptual perspectives, common ground among studies was difficult or impossible to establish, making synthesis intractable. Lack of experimental manipulation and hypothesis testing left explanations of *why* populations changed tantalizingly obscure. Even in 1970 Watson and Moss bemoaned the fact that in vertebrate population studies there was still uncertainty about the major factors involved in limitation or regulation of populations. One reason they gave was studies that were too narrow to permit an estimate of relative importance of factors, and another reason was that "too many studies of vertebrates still rely purely on describing and measuring the natural situation and too few experiments have been done or are being done" (Watson and Moss 1970, p. 209).

However, these limitations with approaches were not to be recognized for many years after the 1950s had passed. And although we can observe a tremendous wave of activity in population ecology building in the 1950s and cresting in the 1960s, apparent advances actually contributed to the concentration of effort using the descriptive paradigm on population dynamics. In the same year, 1954, three very important publications appeared: Lack's book on regulation of animals, Andrewartha and Birch's book on distribution and abundance, and Morris and Miller's paper on the use of life tables for wild populations. These publications did not provide a synthetic approach to population ecology, instead reinforcing the descriptive paradigm, and even molding the kinds of results that could be garnered from such approaches. To establish the validity of this point I will turn to the development of life tables for the study of wild populations.

DEMOGRAPHY AND LIFE TABLES

Morris and Miller's (1954) paper on the development of life tables for the spruce budworm (*Choristoneura fumiferana*), a small tortricid moth whose caterpillars consumed huge areas of coniferous foliage in New Brunswick, Canada, and neighboring provinces and U.S. states, set a standard for entomologists in particular on how to conduct field studies in demography and population dynamics. Therefore, it is worth a brief historical view of how life tables were developed, why they had such impact and set a standard, and why they misguided us, in my view, because the methodology could reveal only one kind of answer. Whether the answer was correct or not probably depended on the kind of species studied, as discussed later in this book.

Demography arose as a discipline because of interest in human populations. Both "population" and "demography" were derived from the Latin and Greek terms for "people" respectively. Cole (1957) provided a detailed view of the development of demography, so I will treat only developments that resulted directly in life tables for wild populations.

The probable remaining length of life for a human has always been an important ingredient for the actuarial profession, at least since Roman times. From empirical studies, survivorship curves were developed for humans (cf. Pearl and Reed 1920; Lotka 1924), a practice adopted for animals by Pearl and Parker (1921) on *Drosophila* and Pearl and Miner (1935) on various invertebrates. A major advance came with Deevey's (1947) application of survivorship curves to natural populations of wild animals. For example, the horns of Dall mountain sheep grow with age, such that a collection of horns from dead animals could reveal the age at death in a population and a survivorship curve.

In addition to survival per unit of time, reproduction could be recorded and the consequent growth rate of the population calculated, as described in any general ecology text book. Hence, the term **life table** was coined.

Soon after Deevey's (1947) paper, Morris and Miller (1954) realized that survivorship curves could be used for the study of insect populations in the field, for long-term population dynamics studies. They started a life table with the cohort of eggs per 10 square feet of branch surface, and with repeated censuses created a survivorship curve for each generation, with the important addition of estimated causes of death for each stage of life censused. The first such tables covered populations of spruce budworm with eggs laid on foliage in 1952 and completion of the generation in 1953 (Table 2.1). One population was relatively low and untreated with insecticide, while another population was more than 10 times higher and sprayed with DDT (Table 2.2). However, as can be seen in the life table, starvation was estimated to be 20 times more effective than insecticide in cause of death in the cohort. Starvation was caused by late frosts, which caused death of new foliage critical to early instar larvae.

We can see in Morris and Miller's life tables an actual record of the causes of death and the demise of the cohort: how the immatures suffered. They are more realistically regarded as **death tables**! When

Table 2.1. *The life table published by Morris and Miller (1954) for a relatively low population of spruce budworm, not sprayed with insecticide, in the Green River watershed, New Brunswick, Canada. The generation started in 1952 and ended in 1953. This was the first life table published on insects in a natural population and the first to include a categorization of mortality factors impinging on each age interval*

Age interval[a], x	Number[b,c] alive at beginning of x, l_x	Factor responsible for d_x, d_xF	Number[b] dying during x, d_x	d_x as percentage of l_x, $100q_x$
Eggs	174	Parasites	3	2[d]
		Predators	15	9[d]
		Other	1	1[d]
		Total	19	11[d]
Instar I	155	Dispersion, etc.	74.40	48
Hibernacula	80.60	Winter	13.70	17
Instar II	66.90	Dispersion, etc.	42.20	63
Instars III–IV	24.70	Parasites	8.92	36
		Disease	0.54	2
		Birds	3.39	14
		Other-inter.[e]	10.57	43
		Total	23.42	95
Pupae	1.28	Parasites	0.10	8
		Predators	0.13	10
		Others	0.23	18
		Total	0.46	36
Moths (sex ratio = 50:50)	0.82	Sex	0	0
Females × 2	0.82	Size	0	0
		Other	0	0
		Total	0	0
"Normal" females × 2	0.82	–	–	–
Generation	–	–	173.18	99.53
Expected eggs	62	Moth migration, etc.	−513	−827
Actual eggs	575			
Index of population trend:	Expected 36% Actual 330%			

[a] Age intervals, x, are defined by developmental stages of the insect that were sampled and are not equal in duration.

[b] Number per 10 square feet of branch surface.

[c] The cohort (ℓ_x) starts with eggs and the mean number per egg mass.

[d] Use of whole numbers results in inaccurate summation for total $100q_x$.

[e] Other factors minus mutual inteference among all factors.

Source: Reproduced from Morris, R. F. and C. A. Miller (1954) The development of life tables for the spruce budworm, *Can. J. Zool.* 32: 283–301 with permission from Natural Resources Canada, Canadian Forest Services, © 1954 Government of Canada.

Table 2.2. *The life table for the 1952–3 generation of spruce budworm at high densities in the Green River watershed, New Brunswick, Canada*

Age interval, x	Number[a] alive at beginning of x, l_x	Factor responsible for d_x, d_xF	Number[a] dying during x, d_x	d_x as percentage of l_x, $100q_x$
Eggs	2176	Parasites	1	<1[b]
		Predators	174	8[b]
		Other	21	1[b]
		Total	196	9[b]
Instar I	1980	Dispersion, etc.	1148	58
Hibernacula	832	Winter	141	17
Instar II[c]	691	Dispersion, etc.	484	70
Instars III–IV[d]	207	Parasites	2.90	1
		Disease	0.30	<1
		Birds	1.70	1
		Starvation	165.30	80
		DDT	8.30	4
		Other-inter.[e]	26.70	13
		Total	205.20	99
Pupae	1.80	Parasites	0.13	7
		Predators	0.11	6
		Others	0.27	15
		Total	0.51	28
Moths (sex ratio = 54:46)	1.29	Sex	0.10	8
Females × 2	1.19	Size	0.57	48
		Other	0.00	0
		Total	0.57	48
"Normal" females × 2	0.62	—	—	—
Generation	—	—	2175.38	99.97
Expected eggs	47	Moth migration, etc.	−199	−423
Actual eggs	246			
Index of population trend:	Expected 2% Actual 11%			

[a] Number per 10 square feet of branch surface.

[b] Use of whole numbers results in inaccurate summation for total $100q_x$.

[c] The "Instar II" cohort sample that estimated loss to a cohort by larvae dispersing on silken threads was completed before spring feeding commenced.

[d] Note the 80% mortality of the cohort in time interval "Instar III–VI" due to late frost killing early foliage. Such loss of food would impact second instar larvae beginning to feed in the spring, but sampling in 1953 was not conducted until larvae had reached the third and fourth instars.

[e] Other factors minus mutual inteference among all factors.

Source: Reproduced from Morris, R. F. and C. A. Miller (1954) The development of life tables for the spruce budworm, *Can. J. Zool.* 32: 283–301 with permission from Natural Resources Canada, Canadian Forest Services, © 1954 Government of Canada.

such tables are developed year after year and reasons are sought to account for population change using correlational methods, and with only mortality factors recorded, clearly one or two mortality factors will turn out to be well correlated with population change. Many authors assumed that such correlation revealed causation, without any independent evidence, making such conclusions a leap of faith. And so, time and time again authors concluded that natural enemies were regulating populations, or competition for food, or weather or disease were regulating. "Death tables" dictate the result to be obtained; that is, some kind of mortality factor will be assumed to cause the population dynamics of the focal species. Such methodology was employed in hundreds of studies and became the mainstay of population dynamics studies on insects, mainly pest species (cf. Harcourt 1969; Cornell and Hawkins 1995). Such analysis was also applied to mammals in some cases, for example East African buffalo (Sinclair 1970) and wildebeest (Sinclair 1973).

The reason why Morris and Miller's methodology became so widely adopted rested on the apparent rigor of the approach, and its conceptual simplicity, even though laborious field sampling was inevitable. Accurate sampling techniques were developed (e.g. Morris 1955, 1960), so long as populations remained high, and key factor analysis developed by Morris (1959, 1963a, b) provided a methodical approach for statistically testing those mortality factors that correlated best with population change from year to year. Introductory coverage of this and other methods was provided in Price (1997).

Following Morris, Varley and Gradwell (1960, 1968, 1970; Varley et al. 1973) developed a similar method, using the winter moth with studies started in 1950, as their organism of choice. Thus, the general method using life tables spread rapidly from Canada to England, involving the two countries most actively involved with research on the population dynamics of insect pests. Influence then spread to Europe and Australia in particular, creating a web of interacting scholars who rose in academic stature to some of the most notable ecologists of their time (cf. Clark et al. 1967; Southwood 1968). Life table analysis and forms of key-factor analysis became entrenched as the major approach to the study of insect populations.

We have listed many of the shortcomings of the life table approach before (Price et al. 1990), so I will introduce them only briefly.

1. The six limitations noted under early field studies continued to prevail and therefore the Idiosyncratic Descriptive Paradigm prevailed.

Cohorts of individuals at the start of the life table were generally eggs, resulting in a large gap in knowledge about female behavior and responses to variable food quality for the young.

2. Regular sampling of a population through a generation was very time consuming and therefore expensive as well. Little time or money was left for experimentation, even if the motivation existed, so that detailed mechanistic empirical studies were absent. Hence, correlational methods were used to interpret causes of population change – a weak and unsatisfactory approach by present-day standards.

3. Plot techniques were generally employed with sampling limited, year after year, to the same designated plots. Spatial dynamics was missed, even if it were readily apparent within plots, because mean values for plots were used, with a gross loss of information on variation. For example, the five oak trees in the studies of winter moth by Varley and Gradwell were dramatically different in phenology and moth population densities (cf. Hunter *et al.* 1997; Hassell *et al.* 1998). Studies on long gradients to reveal spatial variation were virtually absent even though gradients in population density were well known, as in the grey larch tortrix, *Zeiraphera griseana*, in Switzerland (Auer 1961; Baltensweiler 1968).

4. Large samples were taken, usually involving a unit of a plant that provided a standard of leaf area or volume, and mean population density was calculated per plot based on these. Large samples were needed to retain accuracy of population density estimates when densities were low (cf. Morris 1955) and even then many pest species could not be studied between eruptive episodes by life table approaches because individuals became too rare. Crucial studies on the development of new outbreaks were virtually absent in the literature. Populations at their peak and in decline were the principal focus of attention, again predisposing conclusions that involve mortality factors.

All this is easy to note with the advantage of hindsight, but while we espoused the methodology, the results shaped our concepts on population dynamics. Rather than enlightening us further than studies had up to the 1950s, life table approaches tended to entrench earlier views. We may well ask if we have been enlightened since the 1950s and 1960s. I think we have and I will explain my reasons in the next section and in subsequent chapters.

After what I consider to have been the heyday of population dynamics studies in the 1950s and 1960s, two factors resulted in a radical decline in interest. One was the seemingly insoluble debate in those times on density-dependent versus density-independent factors in population regulation, accompanied by an aging population of combatants (cf. Murdoch 1994). The other was the rise of evolutionary ecology in the 1960s with its diverse range of topics that we can now see impinging directly on the population dynamics of organisms.

EVOLUTIONARY ECOLOGY

Collins (1986) has traced the history of the emergence of evolutionary ecology from Lack's (e.g. 1946, 1948a, b, c) views on the evolution of clutch size in birds, to Hutchinson's (e.g. 1965) concern with evolutionary shifts in coexisting guild members, and MacArthur's (1962; MacArthur and Wilson 1967) concepts of r and K selection. Orians (1962) was moved to make a plea for the combination of both functional and evolutionary ecology as a means for avoiding the density-dependent versus density-independent debate. Some quotes from his significant paper are worth repeating because, as far as population dynamics is concerned, they have fallen largely on deaf ears.

> The roots of the current controversy which so deeply divides ecology lie much deeper than their peripheral manifestations in the argument over density-dependence and density-independence. Rather, they stem from the division of the field into two major categories – functional ecology and evolutionary ecology. Both of these approaches are valid and useful and it is a mistake to erect general ecological theory exclusively on either. (p. 262)

> As functional ecologists, Andrewartha and Birch are concerned with the operation and interaction of populations. (p. 260)

> As an evolutionary ecologist, Lack is primarily concerned with the causes behind observable ecological adaptation and has made his major contribution in the subject of the evolution of reproductive rates. This approach leads to the rejection of climate as a significant regulating factor for populations, a rejection which the functional ecologist finds incomprehensible. (p. 260)

> It is pointless to debate the validity of these contrasting approaches to ecology as both have clearly justified their usefulness in all fields of

biology. However, it is of great importance to consider the claim of Andrewartha and Birch that general ecological theory can and should be
built solely upon the functional approach. Just as many physiologists
treat the animal body as a highly interesting and complex mechanism
which has not been and is not going anywhere, Andrewartha and Birch
treat ecology as the study of complex relationships between animal populations and their environments which are best understood as neither
having evolved nor continuing to evolve. (pp. 260–261)

It is becoming increasingly apparent that a complete answer to any question should deal with physiological, adaptational and evolutionary aspects of the problem. The evolutionary process of becoming yields the
most profound understanding of biological systems at all levels of organization. The non-evolutionary answer to the question of why an animal
is abundant in some parts of its range and rare in others is of necessity
incomplete. The functional ecologist can and does make an important
contribution to the understanding of the dynamics of populations, but
for the formulation of theory it is essential that the approaches be combined. The functional approach by itself cannot provide a basis for theory
and, in fact, the "theory" of Andrewartha and Birch really states that no
general theory of ecology is possible and each case must be considered
individually. (p. 261)

Exactly! Hence my labels – the Idiosyncratic Descriptive Paradigm on distribution, abundance, and population dynamics, or the Microecological
Descriptive Paradigm.

Evolution would seem to be the only real theory of ecology today. (Orians
1962, p. 262)

Dobzhansky (1973, p. 129) was equally blunt about the theory of
evolution providing "the golden thread" that runs through and unites
biology. "Seen in the light of evolution, biology is, perhaps, intellec-
tually the most satisfying and inspiring science. Without that light it
becomes a pile of sundry facts – some of them interesting or curious
but making no meaningful picture as a whole." This statement is applicable to most of the history of research on population dynamics, as
it is to many other areas of biology.

In a vein similar to that perceived by Orians, Mayr (1961) emphasized two kinds of explanation for a given phenomenon, one functional and one evolutionary. We can ask why song birds in temperate
zones breed in the spring. The functional answer is physiological, based
upon an understanding of day length and temperature that stimulate
secretion of reproductive hormones. The evolutionary answer is that the

reproductive effort coincides with the highest availability of arthropod food for nestlings, maximizing probable reproductive success. The proximate explanation is physiological (the cause of breeding time) and the ultimate explanation is evolutionary (the adaptive effect of breeding time). Both are essential to a full mechanistic understanding of breeding time in song birds, as Orians explained.

Given these emphases on evolutionary thinking in the early 1960s during the full flush of interest in population dynamics and following the centennial of the publication of *The Origin*, we must wonder why so many persisted in a microecological mode, especially entomologists. Those studying other taxa had espoused evolutionary thinking by the early 1960s (e.g. Chitty 1957, 1960, 1967; Wynne-Edwards 1962, 1964, 1965) even if imperfections in their logic became apparent. Part of the problem undoubtedly derives from ecologists, even today, forgetting the power of evolutionary theory in comparative studies in any biological field. This power was understood by Darwin (1859) and expressed by him most clearly in the last paragraph of his book (pp. 489–490).

> It is interesting to contemplate an entangled bank, clothed with many plants of many kinds, with birds singing on the bushes, with various insects flitting about, and with worms crawling through the damp earth, and to reflect that these elaborately constructed forms, so different from each other, and dependent on each other in so complex a manner, have all been produced by laws acting around us. These laws, taken in the largest sense, being Growth with Reproduction; Inheritance which is almost implied by reproduction; Variability from the indirect and direct action of the external conditions of life, and from use and disuse; a Ratio of Increase so high as to lead to a Struggle for Life, and as a consequence to Natural Selection; entailing Divergence of Character and the Extinction of less-improved forms. Thus, from the war of nature, from famine and death, the most exalted object which we are capable of conceiving, namely, the production of higher animals, directly follows. There is grandeur in this view of life, with its several powers, having been originally breathed into a few forms or into one; and that, whilst this planet has gone cycling on according to the fixed law of gravity, from so simple a beginning endless forms most beautiful and most wonderful have been, and are being evolved.

Darwin realized that five laws could account for the unity and diversity of life. Although the concepts were simple, they explained mechanistically how and why all organisms were related and how and why such strong patterns in phylogeny, life history, behavior, physiology, morphology, and anatomy existed. Darwin provided the strongest possible basis for comparative biology because his view was comprehensive, covering all organisms on earth, and therefore the comparative method was unlimited in its scope. Paradoxically, the largest taxon of organisms, the insects, providing the broadest basis for comparative studies, was not exploited by those interested in population dynamics in an evolutionary and comparative manner.

That those working on insect population dynamics used purely ecological approaches is not so difficult to explain, especially for one who worked through a degree in forestry, worked in forest entomology research, and studied aspects of insect population dynamics in the 1960s. First, academic education for agriculturalists and foresters commonly diminished or excluded any kind of evolutionary education. Second, solving pragmatic problems of insect control tended to focus attention on the ecology and immediate factors, rather than "esoteric" views on apparently abstract subjects like evolution. Third, as I have pointed out, the rush to sample and the use of mean population size milked dry the energy for experimental methods and an emphasis on variation in the population. Descriptive dynamics was the common mode of enquiry, an approach now explicitly rejected in the editorial policy of the journal *Functional Ecology*. The evolutionary synthesis was never espoused, and population dynamics has yet to be enfolded into that synthesis.

Therefore, it is interesting to enquire how the field of population dynamics has changed and why it has changed. My view is that the use of evolutionary ecology provides much of the answer. First, it provided answers to "why" questions. Why does this insect feed on that plant and not on another? Multiple questions of this kind promoted the rapid increase of interest in plant and animal interactions involving plants, herbivores, pollinators, and frugivores and fruit dispersers. There existed already a long tradition of studies on plants, insects, and natural enemies in the agricultural and forestry literature, but now a more evolutionary perspective emerged. Second, an emphasis on experimental approaches and mechanistic answers prevailed, such that both proximate and ultimate answers became available. Third, as technology advanced, so did chemical ecology, a field that provided many of the proximate answers to questions. Fourth, life history evolution became a major theme with its very strong comparative approaches. Fifth, behavioral ecology appears to have been adopted more intimately into other aspects of evolutionary ecology, bringing with it its long tradition of evolutionary thinking and comparative studies, since the early influences of Lorenz and Tinbergen. The move into foraging behavior with

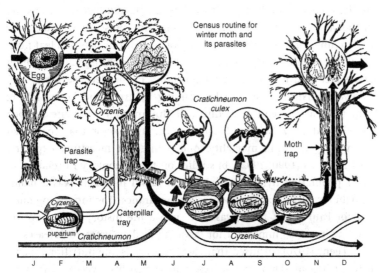

Fig. 2.3. The life cycle of winter moth, *Operophtera brumata*, illustrated by
Varley (1971). Eggs are laid in the tree canopy by flightless females in
November and December. Larvae hatch in April, feed rapidly, and fall to
the ground in late May where they pupate, with pupae spending the
summer and fall in the ground. *Cyzenis* is a tachinid fly parasitoid that
attacks larvae in the tree canopy and *Cratichneumon* is an ichneumonid
wasp parasitoid attacking pupae in the ground. Falling larvae were
caught in sampling boxes in the spring, and climbing females were
trapped late in the year, both to estimate densities for life table
construction. (From Varley, G. C., G. R. Gradwell, and M. P. Hassell (1973)
Insect population ecology, Fig. 7.1, Blackwell Science, Oxford.)

its consequences for population ecology provided a strong mechanistic
approach (e.g. Hassell and Varley 1969; Hassell 1970; van Alphen and
Vet 1986; Godfray 1994).

Let me provide one example of how some of these elements of
evolutionary biology blended into a more complete explanation of nature, in the manner championed by Orians. The example also enables
us to pick up the chain of influence from early field studies on German
insects by Schwerdtfeger, which interested Varley so much, which probably stimulated the initiation of his 19-year study of the winter moth,
and which subsequently provided Feeny (1970) with an interesting
question. The life history of the winter moth, illustrated by Varley (1971)
(Fig. 2.3), involves flightless females climbing trees in November and
December in north temperate England. They lay eggs that hatch early
in the spring coincident with the flushing of early oak foliage. Clearly,

the life cycle had evolved in a way that maximized first instar feeding on the youngest possible foliage. The selection pressure was sufficient on hatching early that the probability of asynchrony between leaf flush and larval eclosion was real. In years of some asynchrony early-hatching larvae died and became a major mortality factor. Because Varley and Gradwell censused only females walking up trees and larvae falling from trees at the end of feeding, all this mortality became lumped into the category "winter disappearance," the factor that correlated best with generation mortality. And, while not measured explicitly until Hunter's (1990, 1992a, b) studies on the phenology in relation to insect population dynamics, first instar larval mortality was a major contributor to the so-called winter disappearance. The interesting question for Paul Feeny, then, was why such a life history should evolve, and why it was so important for larvae to feed on young oak leaves.

Feeny (1970) surmised that there must be some nutritional benefit to feeding on young oak leaves, constructed his own gas chromatograph before they were commercially available, and proceeded to analyze chemicals in oak leaves throughout their lives on this deciduous tree. He noted a rapid decline in feeding caterpillars in May and examined sugars, water, protein, and tannin contents in leaves. Because tannins are known to combine with proteins, making digestion of proteins largely impossible, he also estimated the protein/tannin ratio to provide an index of the protein available for larvae. Feeny discovered that water content declined with leaf age, as did protein and the protein/tannin ratio, with increasing tannin content exacerbating the shortage of protein in a diet already low in protein (Fig. 2.4). Clearly, then, the ultimate factor selecting for the evolution of the winter moth life cycle, with eclosion early in the spring, was high nutritional leaf content and rapid larval development, possible in the early spring. The proximate explanation for early hatching is increasing temperatures that trigger embryonic development, the details of which have not been elucidated for this species.

This mechanistic explanation for the evolution of early feeding, coupled with the potential for high mortality when a mismatch between eclosion and leaf flush occurred, also contributed to an understanding of a general pattern. On oaks, eruptive and very abundant species such as winter moth and green tortrix feed early on the youngest foliage, develop rapidly, and then pupate early before tannins are highly concentrated (Fig. 2.5). Other moths that feed later in the season, when tannin content is high, develop slowly, and are much less eruptive in their population dynamics.

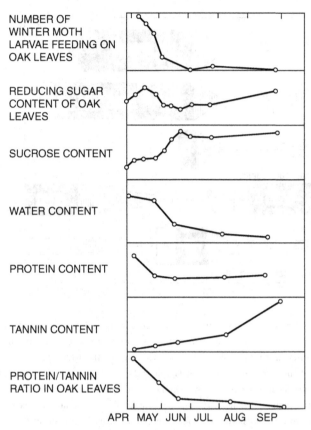

NUMBER OF WINTER MOTH LARVAE FEEDING ON OAK LEAVES

REDUCING SUGAR CONTENT OF OAK LEAVES

SUCROSE CONTENT

WATER CONTENT

PROTEIN CONTENT

TANNIN CONTENT

PROTEIN/TANNIN RATIO IN OAK LEAVES

APR MAY JUN JUL AUG SEP

Fig. 2.4. Feeny's (1970) results on oak leaf constituents in young leaves in April to old leaves in September. Two variables correlate well with decline in winter moth numbers in trees: water content and protein content. Because larvae of other species can feed later in the season (cf. Fig. 2.5), water content appears not to be limiting, whereas available protein, indicated by the protein/tannin ratio, declines rapidly while larval numbers decline.

Feeny's example illustrates the kind of explanatory power in an evolutionary ecology approach. We gain an explanation of the evolution of a life history that was not intuitively obvious, and yet exceedingly common among species with leaf-feeding caterpillars on woody plants, including the moths and butterflies, sawflies, leaf beetles, and others. We see chemical ecology playing its role in the understanding of plant and herbivore interactions and light is cast even on possible three-trophic-level interactions involving plants, insect herbivores, and natural enemies. For if tannins reduce the growth of insect herbivores

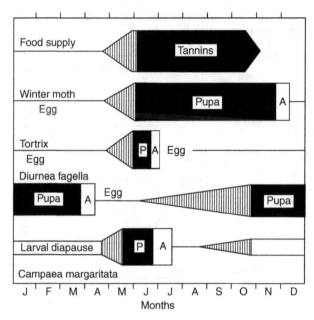

Fig. 2.5. The relationships among food supply, tannin content of oak leaves, and feeding patterns of moths on oak: the early feeders, the winter moth, *Operophtera brumata*; the green oak tortrix, *Tortrix viridana*; and the late feeders, *Diurnea fagella* and *Campaea margaritata*. Note the short feeding time of early feeders when tannins in leaves are low and the much longer feeding times of late-season caterpillars on leaves high in tannins. *Campaea* larvae feed for two months in late summer, developing slowly, and for another month in April and May, developing rapidly when leaves are young. Timing is for a north temperate climate in Wytham Wood, a property of Oxford University. (From Varley 1967.)

without killing them, caterpillars are likely to feed more and inflict more damage on the plant, a paradox for those who invoke tannins as plant defenses against being eaten. But Feeny (1975, 1976) argued that delayed larval development would increase exposure to natural enemies, resulting in plants benefiting from nonlethal defenses (see also Price *et al.* 1980).

Further progress was made in seeking patterns in nature using evolutionary thinking when Feeny moved from Oxford University to Cornell University in 1966 and started to work on the cabbage family (Brassicaceae), or crucifer species of plants and their herbivores. Comparing the chemical ecology of oaks with the herbaceous crucifers, and contrasting life histories of long-lived trees and short-lived annual or biennial herbaceous plants revealed very broad patterns in nature, as explained in Feeny's (1975, 1976) **Plant Apparency Hypothesis**. Very briefly, long-lived trees which are bound to be found by herbivores invest heavily in costly chemical defenses with broad-spectrum efficacy. These quantitative defenses are expensive but the cost is tolerable for a long-lived plant. Short-lived plants are less easily detected by herbivores, and their best defense is being hard to find in patchy and ephemeral sites. Low-cost defenses are effective against generalist herbivores, should plants be found. Instead of tannins and other digestibility reducers found as defenses in long-lived plants, short-lived plants have evolved with mustard oils (glucosinolates) in crucifers, for example, alkaloids in the potato family, and furanocoumarins in the carrot family.

This evolutionary and ecological approach enables broad patterns to be detected with a mechanistic explanation and thus provides the basis for theory in ecology. While the Plant Apparency Hypothesis has been criticized by some (e.g. Coley 1983), my view is that it remains a compelling insight into plant–herbivore interactions with the broadest explanatory power of any hypothesis we have in the field. A revealing exercise that Paul Feeny used to try on his ecology classes is to ask what students eat in the plant kingdom and what they do not eat. On one axis of a matrix he ordered plants in ecological succession from ephemeral weeds to perennial herbs, to shrubs and trees, including the plants we eat derived from such wild plants. On the other axis he listed the parts of plants: roots, shoots, leaves, flowers, seeds, fruits. The green vegetables are all derived from herbaceous plants in early succession: weedy species in the wild. But we do not eat the green leaves of trees. Root crops such as carrots, parsnips, radishes, beets, and turnips are all derived from annual or biennial plants, but tree roots are not on the menu. The only produce from trees that we commonly eat are fruits and seeds (and the rare pseudofruits of such plants as cashews, *Anacardium occidentale*, with its "cashew-apple"). The patterns of what we do and do not eat of the vegetable kingdom fall out very clearly, and are fully consistent with the Plant Apparency Hypothesis.

The Plant Apparency Hypothesis is characteristic of some hypotheses on plant–herbivore interactions in being pluralistic; it explains why some plants are heavily defended by digestibility reducing compounds and others are lightly defended by toxic chemicals. When the starting points are fundamentally different, such as trees and weedy species, the evolutionary consequences are divergent. Other examples include the Carbon-Nutrient Balance Hypothesis and the Resource Availability Hypothesis (cf. Price 1997). Because of their pluralistic nature they provide more insight and cover more ground than singular hypotheses. Other hypotheses, such as those reviewed by Price (1997) on plant–herbivore interactions, include the Plant Stress Hypothesis, the Induced Defense Hypothesis, and the Plant Vigor Hypothesis. Each of these hypotheses has attempted to be general, without being pluralistic and therefore setting limits on a certain kind of interaction. Without the ability to state in which circumstances the hypotheses will apply and when it will not apply, the detection of broad patterns in nature is left in a difficult position. We tend to collect more and more studies that are consistent or inconsistent with a particular hypothesis without an ability of prediction and without much insight into categorization into classes of interactions. This is a generally valid criticism, I think, in large areas involving plant and herbivore interactions such as chemical ecology, multitrophic level interactions, induced defenses in plants, and herbivore–carnivore interactions. We have many examples, most of them interesting and insightful, but we have yet to find a means whereby broad synthesis and theory can be developed. For example, in their excellent book on induced defenses of plants, Karban and Baldwin (1997) reviewed the literature but provided very few generalizations, patterns, or predictions on where such interactions should or should not be found in nature, and why they evolved based on plant fitness studies.

The field of evolutionary ecology produced tremendous research energy in the 1970s to the present day, and plant–animal interactions became one of the most vigorous fields. We learned many important lessons relevant to population dynamics of herbivores. Chemical ecology showed us that green plants were highly variable in time and space and much of the greenery was toxic or inadequate, especially in nitrogen. Phytochemical aspects of the plant–herbivore interaction developed rapidly (e.g. Rosenthal and Janzen 1979; Rosenthal and Berenbaum 1991, 1992). We noted that trophic levels based on terrestrial plants were intimately connected through physical and chemical mechanisms (e.g. Price et al. 1980) and that body odors of plants and animals were a major modality of communication and detection (e.g. Nordlund et al. 1981). Variation in species and interactions was emphasized, as a basis for natural selection, rather than using mean values to describe plant qualities or population densities (e.g. Denno and McClure 1983), and resource variation in plants for herbivores focused attention on bottom-up influences on distribution, abundance, and population dynamics (e.g. Hunter and Price 1992a; Hunter et al. 1992).

Thus, when renewed interest in population dynamics in wild populations emerged in the 1980s, the necessary components of the system to be investigated had increased considerably, based on studies in evolutionary ecology. We were better prepared than in the 1950s and 1960s to examine population phenomena using both observational and experimental approaches. Importantly, an evolutionary perspective was developing. I would like to recount some of the developments that I think of as important in the development of evolutionary aspects of population dynamics.

AN EVOLUTIONARY PERSPECTIVE ON POPULATION DYNAMICS

While working on spruce budworm population dynamics Morris realized the constraints imposed by plot techniques and the necessity of frequent and grueling sampling. He realized the need for experimental approaches, which he called process studies (Morris 1969), or mechanistic studies, and the advantage of extensive geographic population surveys to search for large patterns in distribution. He selected for study the fall webworm, *Hyphantria cunea*, whose larvae spin a conspicuous web that can be censused readily while driving along roads. His research, published between 1964 and 1976, was summarized by him in 1969 and I have provided an overview of his work with subsequent papers cited (Price 1997).

Morris (1969) noted ten ways in which heat available for the development of the fall webworm influenced population dynamics, and heat in terms of degree-days for development took a central place in his modeling. He showed an immediate impact of heat on one generation and delayed effects of heat from one generation to another. He found that heat also affects the food quality of foliage with an effect on the current generation, and on the fecundity of females, with delayed effects in the subsequent generation. He showed that there was heritability for heat requirements and that natural selection on heat requirements caused populations to change rapidly from year to year as degree-days per year changed. Heat also affected the influence of parasitoids and the efficacy of larval defense against internal parasitoids. He also explained the differences in geographic variation in population response to cold weather based on the evolution of populations under typical temperature conditions in each climatic zone.

Morris fully incorporated the importance of geographic variation, natural selection, and heritability into his studies and modeling of the fall webworm. His was extraordinarily effective research on population dynamics, and an example to us all. However, nobody seems to have followed his example. He worked in a Canadian forest research laboratory, without directing doctoral students as in a university setting, and became reclusive, so that his influence was diminished.

Morris concluded his 1969 paper with the following statement, touching on several important components of his model and hopes for the future.

> The model should represent a higher level of biological meaning than could be achieved through regression analysis based on field data alone. My hope is that it will be good enough for reliable simulation studies, with the object of learning whether or not density dependence represents an essential aspect of the webworm's system of regulation and what would happen if sequences of warm years extended beyond their normal expectancy. As a result of the effects of natural selection on heat requirements, webworm populations that are increasing during a series of warm years become progressively less able to take advantage of these favorable conditions (i.e., they need more and more heat for the same amount of development). It will be instructive to learn how much the genetic parameters in the model, by themselves, contribute to population stability. Finally, it can be shown that population density is related to land use, vegetation types, and other variables...It will, therefore, be worthwhile to employ simulation and minimization techniques to see whether cultural manipulation of the environment can be used feasibly to reduce webworm damage. (p. 27)

He noted the improved explanatory power of his process studies compared with field studies alone and correlational interpretations. He was also examining the role of density dependence, the genetic quality of populations, and intrinsic mechanisms in population regulation. Host-plant phenology and quality variation were constituents of the model, as was the geographic variation of land use and climate. Morris's legacy is a vastly more comprehensive enquiry into the distribution, abundance, and population dynamics of a species than had been achieved up to his time and remains to this day as an example we can all benefit from.

Quite independently from Morris's approach, Southwood (1975) and Southwood and Comins (1976) developed a population dynamics model explicitly based on evolutionary ecology (Fig. 2.6). The argument was based on r- and K-selection as representing life history traits of insect species, with r-selected species in unstable environments such as agricultural fields and K-selected species in stable habitats, as in temperate forests. Intermediate between extreme r-selected species and extreme K-selected species was a stable equilibrium, reached when populations rose to a density at which natural enemies counteracted

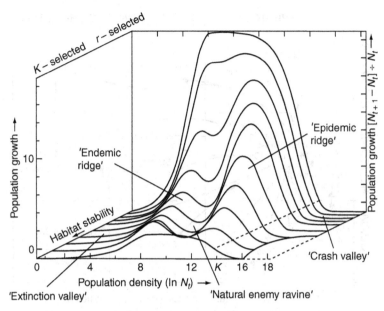

Fig. 2.6. A model of population dynamics by Southwood (1975) and Southwood and Comins (1976) involving habitat stability from low to high, species from r-selected to K-selected life histories respectively and the effects of natural enemies creating a "natural enemy ravine" and depletion of resources at high densities resulting in a precipitous decline into "crash valley." (From Southwood, T. R. E. and H. N. Comins (1976) A synoptic population model, Fig. 1, J. Anim. Ecol. 45: 949–965, Blackwell Science, Oxford.)

population growth, establishing a "natural enemy ravine" and holding populations at an "endemic ridge." Should populations escape the role of natural enemies an "epidemic ridge" would develop, probably followed by a crash to low populations through starvation.

This model for the first time, I think, brought life history differences in species into full play in a population dynamics scenario. Unfortunately, it is too simple to capture the range of dynamics observed in nature, and is clearly not supported by the many eruptive species in temperate forests. However, an important germ was planted into thinking about population dynamics: that life history differences may play an important role in the development of different dynamical patterns. To what extent Southwood's ideas influenced subsequent workers I have been unable to determine, but certainly a life history approach was adopted by several researchers a decade later, and his work is frequently cited by Wallner (1987), discussed below.

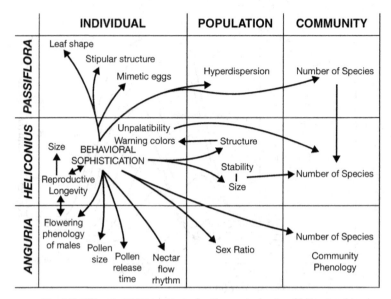

Fig. 2.7. Gilbert's (1975) scenario for the central role of behavioral sophistication among *Heliconius* butterflies, with its strong effects on population structure, stability, and size and community organization of *Heliconius* species. The *Heliconius* butterflies also influence individual, population, and community characteristics in the larval food plants, *Passiflora* (Passifloraceae) species, and the adult food plants, *Anguria* (Cucurbitaceae), which provide pollen and nectar to the butterflies. (From *Coevolution of animals and plants*, eds. L. E. Gilbert and P. H. Raven, © 1975, 1980, by permission of the University of Texas Press.)

An illuminating and novel approach to understanding population phenomena was developed in the early 1970s by Gilbert (1975, 1991), who emphasized the evolution of behavioral traits as a starting-point (Fig. 2.7). *Heliconius* butterflies evolved behavioral sophistication in traplining for nectar and pollen, with a capability for pollen digestion. This resource provided adults with enough nutrients for a long life in which learning the local habitat and passing knowledge from generation to generation became adaptive. Both the *Passiflora* food plants for larvae and *Anguria* pollen and nectar sources for adults were widely distributed (hyperdispersed), which could only be exploited by species with the behavioral sophistication identified by Gilbert. Various life history traits contributed to the structure of populations, their size and stability with the key evolved behavioral traits as central. While I do not believe that Gilbert's studies influenced those working on population dynamics, his example illustrates nicely how life history traits and behavior can be used as effective evolutionary bases for understanding populations, and indeed such an evolutionary approach may well be essential. That the kind of ecology observed in a species is a direct consequence of its evolved traits was the message that Gilbert (1975) conveyed in the title of his paper.

The decade of the 1980s, I regard as a reawakening of much interest in population dynamics, much of the activity focusing on life history traits correlated to outbreak dynamics. At this time we began to see sufficient activity so that researchers were directly and rapidly influencing each other's approaches. Themes emerged emphasizing the evolved characteristics of species that correlated with outbreak or nonoutbreak dynamics: fecundity, clutch size, gregarious feeding, specificity in food and habitat, and so on.

Wallner (1987) used a gradient from r-selected to K-selected species, as had Southwood, and ranked the relative importance of mostly life history traits in relation to rare and irruptive insect species (Table 2.3). This approach of searching for correlates of evolved characters with population dynamics was also used by Nothnagle and Schultz (1987), Barbosa et al. (1989), Hunter (1991), Haack and Mattson (1993), and Larsson et al. (1993). Frequently, gregarious feeding correlated well with outbreak species, but why gregarious feeding evolved was not explained. A fully mechanistic explanation remained elusive. A very revealing approach was developed by Barbosa et al. (1989) and Hunter (1995a, b) that explained the evolution of outbreak species of forest moths and the common occurrence of flightlessness in females of the species. I have tried to summarize the scenario developed (Price 1997) (Fig. 2.8), starting with conditions typical of north temperate forests, moving to evolutionary responses of moths in these forests, and on to the ecology in which we observe eruptive population dynamics. The explanation is clearly evolutionary in nature. And colonial feeding is a character *derived* from poor mobility of females in this scenario, and not a mechanistic explanation for eruptive dynamics: it is circumstantially correlated with outbreak species.

During these years of relatively intense publication activity my own research group was developing scenarios related to the Phylogenetic Constraints Hypothesis (Price et al. 1990, 1995a; Price 1992b, 1994b). Evolved traits of species were used as the basis for detecting and explaining patterns in the distribution, abundance, and population dynamics of insect species.

Coincident with these developments was the publication of several edited and single-author volumes on insect population dynamics

Table 2.3. *Evaluation of biological traits associated with different kinds of insect population dynamics. The relative importance of a trait increases with the number of plus signs indicated*

| Biological trait | K-selected | | | r-selected |
	Rare[a]	Gradient[b,c]	Cyclic[b,d]	Irruptive[b,e]
Adult life span	+++	+++	+	+
Adult feeding	+++	+++	+	+
Habitat restriction	+++	+++	+	+
Response to plant defense	+++	+	+	++
Degree of host specificity	+++	+++	+	++
Flush–crash cycles	+	+++	+++	+++
Adult vagility	+	++	++	+++
Fecundity	+	++	+++	+++
Alteration of environment	+	+	++	+++
Degree of egg clumping	+	+	+++	+++
Degree of larval aggregation	+	+	+++	++
Response to weather	+	++	+++	+++
Importance of biological control	+	++	+++	+++
Utilization of foci or refuges	+	+	+++	+++
Incidence of polymorphism	+	+	+++	+++

[a] Rare species persist at low densities.

[b] Outbreak types of gradient, cyclic, and irruptive are based on categories defined by Berryman and Stark (1985).

[c] Pest gradients occur when site conditions change, such as drought, favoring insect population increase.

[d] Pest cycles result from intrinsic factors which result in predictable cycles.

[e] Pest irruptions are defined as irregularly spaced flushes to very high densities interspersed with very low densities.

Source: Wallner (1987). Reproduced, with permission, from the *Annual Review of Entomology*, vol. 32, 1987, by Annual Reviews Inc.

in the field: Barbosa and Schultz (1987) on insect outbreaks; Watt *et al.* (1990) on forest insects; Cappuccino and Price (1995), Dempster and McLean (1998), and Berryman (1999) on various arthropod species; a veritable cornucopia of books appeared compared to the decade before 1987.

Certainly in the late 1980s and in the 1990s evolutionary approaches to population dynamics were gaining ground relative to purely ecological approaches. (1) Many investigators started with life history traits or behavior in order to understand population phenomena. (2) A strong comparative approach also developed, taking in a bigger picture

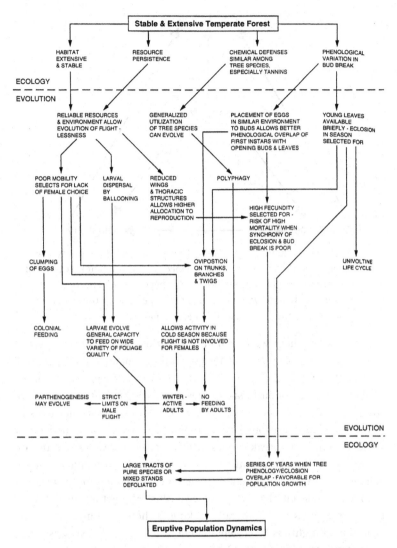

Fig. 2.8. A summary flow-diagram of factors in temperate forest that contribute to the evolution of flightless moths (Macrolepidoptera) in these habitats and the consequent eruptive population dynamics observed. (From Price, P. W. (1997) *Insect ecology*, 3rd edn, © 1997, reprinted by permission of John Wiley & Sons, Inc.)

of all or many outbreak species to examine which traits they held in common. (3) Certainly this **macroecology** of population dynamics, involving the comparison of many species rather than the **microecology** concerned with the details of one, was a healthy development. (4) It placed the field in a strong pattern-finding mode.

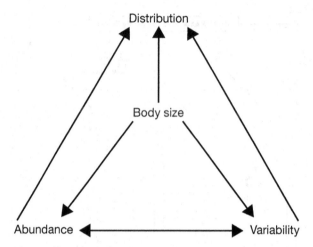

Fig. 2.9. Hypothetical and empirical relationships between body size as it relates to the distribution, abundance, and population variability of organisms. Arrows indicate direction of hypothesized effects. (From Gaston and Lawton (1988a) reprinted with permission from *Nature* 331: 709-712, Fig. 1 © 1988 Macmillan Magazines Ltd.)

However, there were limitations to this general move to evolutionary arguments to account for population dynamics. (1) Papers concentrated on pest species and outbreak species and generally ignored rare or so-called nonoutbreak species. We knew very little about such species. (2) The approaches were more correlational than mechanistic. Thus, although gregarious feeding repeatedly emerged as a moderately frequent trait in outbreak species, why it evolved and why it should correlate with outbreak dynamics was not clarified. (3) Although a macroecological approach to population dynamics was emerging, it was not founded on a strong base of macroevolution. Big picture, macroevolutionary scenarios were generally lacking, although arguments on the evolution of flightless Macrolepidoptera in temperate forests was a very promising entry into the field.

MACROECOLOGY

The emergence of macroecology in the 1980s resulted from broad pattern detection in rich data sets on body size, abundance, geographic distribution, and variation in abundance (e.g. Brown 1981, 1984, 1995; Gaston 1988; Gaston and Lawton 1988a, b; Brown and Maurer 1989; Lawton 1990, 1991). Focus was precisely on the central issues in ecology – distribution, abundance, and population dynamics (Fig. 2.9) – with a variety of taxa showing strong empirical patterns: for example correlations between abundance and geographic distribution, variability and distribution, and variability and abundance (Fig. 2.10). Surprisingly, body size correlated poorly with other variables except for bracken herbivores (Gaston and Lawton 1988a), but for other combinations correlation coefficients were remarkably high (e.g. $r = 0.782$ for 97 species of winged aphids in the abundance–distribution correlation), the range of coefficients was from $r = 0.463$ to 0.782 for all tests on the abundance–distribution relationship on all moths, noctuids, geometrids, aphids, carabids, and bracken fern herbivores. Coupled with similar approaches on mammals and birds by Brown and Maurer (1989) and Brown (1995) major advances were made on large-scale pattern detection using empirical data.

These were the kinds of patterns so important in the development of theory. We can detect the enthusiasm in early papers that discuss the development of "simple empirical and theoretical rules linking population dynamics, distribution and body size." "The generality of these patterns ... may lead to a general theory of animal population biology and to an understanding of evolutionary constraints acting on entire species assemblages"(Gaston and Lawton 1988a, p. 711). And from Brown and Maurer (1989, p. 1149), "Our analyses suggest that the ecological and evolutionary processes that determine the assembly of continental mammal and bird faunas are reflected in regular patterns of body sizes and geographic range configurations. Comparisons of these patterns across spatial scales suggest mechanistic hypotheses that appear to be supported by available data."

Making the mechanistic explanation for these patterns is certainly a challenge (Root 1996; Gaston and Blackburn 1999). Experiments are difficult on large-scale phenomena, and very careful observational data and modeling are probably required to test multiple alternative hypotheses. However, at a time of seemingly general preference for strongly reductionist ecology, the development of macroecology was most heartening for those of us who value synthesis and theory in ecology (cf. Lawton 1991, 1999; Brown 1995; Gaston and Blackburn 1999).

PHYLOGENIES, BEHAVIORS, AND LIFE HISTORIES

Other trends were developing in the 1990s, outside of insect population dynamics studies, which involved pattern detection and mechanistic explanations based on evolved life history and behavioral traits. Harvey and Pagel (1991) published their important book on *The comparative*

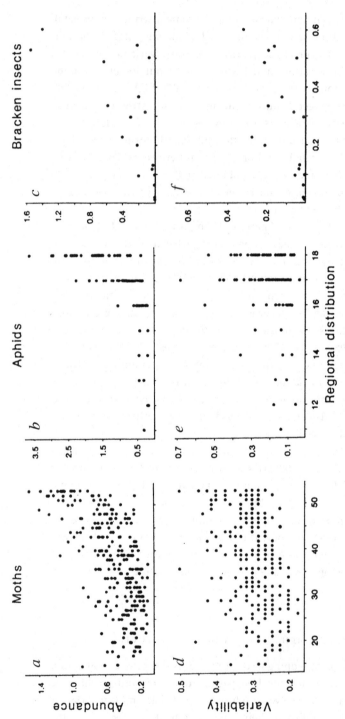

Fig. 2.10. Empirical results on relationships among local abundance and regional distributions for moths (*a*), aphids (*b*), and bracken fern insects (*c*), and population variability and regional distributions for the same taxa (*d–f*). Data are for 263 species of adult British moths, 97 species of winged aphids, and 21 species of insects on bracken fern. All relationships are positive and all are significant except in *f*. (From Gaston and Lawton (1988a) reprinted with permission from *Nature* 331: 709–712, Fig. 2 © 1988 Macmillan Magazines Ltd.)

method in evolutionary biology, opening the book with simple but important truths: "Distantly related organisms work in similar environments and have often evolved similar adaptations to help them in their tasks. Viewed in this way, there are repeated patterns of evolutionary change. If evolutionary biologists think that some trait of interest may have evolved to do a particular job, they naturally use the comparative method when they ask if other organisms doing the same job have evolved similar traits" (p. v). A strongly comparative emphasis in behavioral studies has a long tradition, but a growing realization has emerged that behavior may be central in integrative biology, as Real (1992, p. S1) pointed out so clearly: "Ecological phenomena and community organization can be viewed, to a large degree, as the immediate consequence of individual actions and behaviors." He argued that behavior depends upon internal processes, molecular, cellular, and physiological, and behavior affects external phenomena concerning population and community characteristics. Therefore, behavior may well form the basis for unifying biological sciences. These kinds of approaches in biology have flourished, represented in edited works such as *Phylogenies and the comparative method in animal behavior* (Martins 1996), *The evolution of mating systems in insects and arachnids* (Choe and Crespi 1997a), and *The evolution of social behavior in insects and arachnids* (Choe and Crespi 1997b). And behavior was emphasized in population dynamics as a critical mediator of bottom-up and top-down effects on herbivores by Zwölfer and Völkl (1997).

The comparative method, focusing on behavioral traits, was brought to bear upon the evolutionary basis of population regulation in mammals by Wolff (1997). The approach was similar to the one advocated in this book and earlier by our research group (Price *et al.* 1990), although the results are different. Wolff searched for the basic behavioral and life history traits that produced inevitable consequences for population regulation. That is, he employed evolved characters in a mechanistic explanation for different types of regulation or emergent properties. He contrasted altricial and precocial young, female behaviors in response to developmental stage at birth of young, and grouped taxa according to the convergent patterns observed in nature (Fig. 2.11). Then he developed evolutionary paths that lead to ecological consequences in population regulation: territorial species were regulated intrinsically and nonterritorial species extrinsically (Fig. 2.12). The patterns show much phylogenetic consistency, for example most rodents are territorial, but also much convergence, subjects of concern when dealing with insects also.

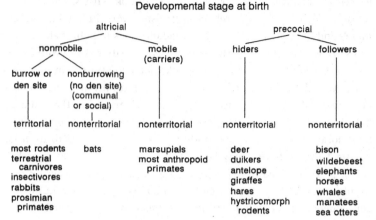

Fig. 2.11. Wolff's concept on the fundamental importance of developmental stage at birth, altricial or precocial, and the mobility of progeny in the convergence of mammal groups into territorial or nonterritorial types of social organization. Note that patterns are to some extent independent of food type and phylogeny. (From Wolff, J. O. (1997) Population regulation in mammals, Fig. 1, *J. Anim. Ecol.* 66: 1–13, Blackwell Science, Oxford.)

A focus on life histories and behaviors as evolved traits that impact ecology directly promises a rich future in comparative studies, in both vertebrates and invertebrates. And a coupling of evolutionary and ecological approaches to distribution, abundance, and population dynamics is essential, as Orians (1962) emphasized so long ago. There appears to be a convergence in emphasis from different independent sources on an evolutionary understanding of life history and behavior as basic kinds of natural history essential in accounting for many ecological consequences (e.g. Ligon 1993; Balda *et al.* 1996; Promislow 1996). As Ricklefs (2000b, p. 3) has said, "The study of life histories today is an active, multifaceted program of research that unites behavior, ecology, population biology, and evolution into a broad concept of the responses of organisms and populations to the conditions of their environments." But, the discovery of pattern and mechanism in ecology is still in its infancy, with great potential for rapid advances as we bring population dynamics fully into the bosom of evolutionary biology.

All this history has influenced my views on the distribution, abundance, and population dynamics of animals, but none of this has been as crucial in the development of my thesis as empirical studies on

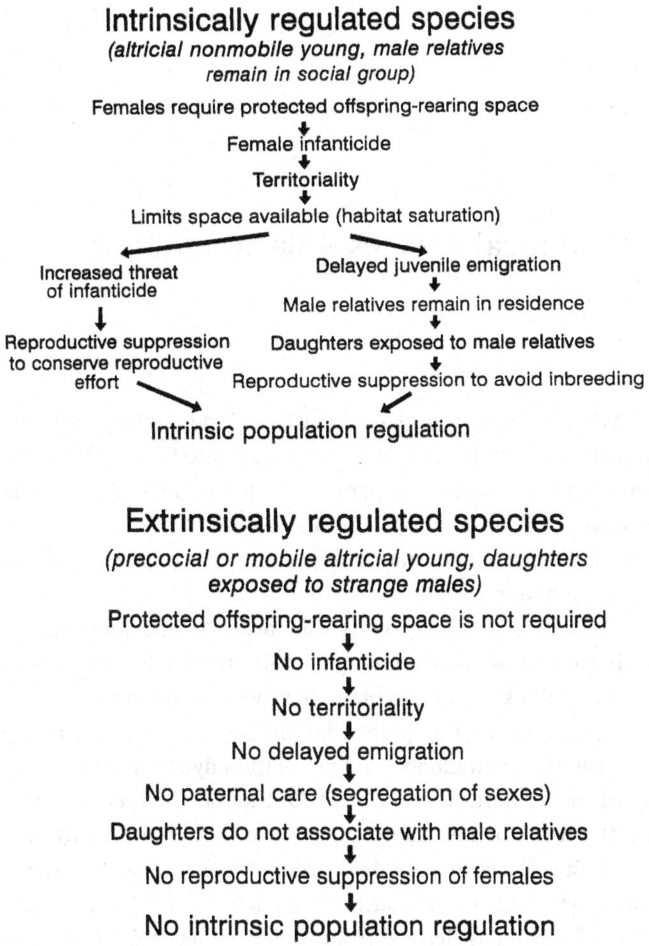

Intrinsically regulated species
(altricial nonmobile young, male relatives remain in social group)

Females require protected offspring-rearing space
↓
Female infanticide
↓
Territoriality
↓
Limits space available (habitat saturation)

Increased threat of infanticide   Delayed juvenile emigration
↓   ↓
Reproductive suppression   Male relatives remain in residence
to conserve reproductive   ↓
effort   Daughters exposed to male relatives
                                                    ↓
                                  Reproductive suppression to avoid inbreeding

Intrinsic population regulation

Extrinsically regulated species
(precocial or mobile altricial young, daughters exposed to strange males)

Protected offspring-rearing space is not required
↓
No infanticide
↓
No territoriality
↓
No delayed emigration
↓
No paternal care (segregation of sexes)
↓
Daughters do not associate with male relatives
↓
No reproductive suppression of females
↓
No intrinsic population regulation

Fig. 2.12. Evolutionary pathways for intrinsic and extrinsic regulation of mammal species, according to Wolff (1997). (From Wolff, J. O. (1997) Population regulation in mammals, Fig. 4, *J. Anim. Ecol.* 66: 1–13, Blackwell Science, Oxford.)

one little sawfly. The following chapters recount the genesis of an idea and a viewpoint that has shaped the historical perspective presented in this chapter and what I perceive as the essential macroevolutionary approach to these central areas of ecology.

3

The focal species – Basic biology

When starting a research program in population ecology, a narrow focus is desirable. If we can understand one focal species well, then it will form the basis for comparative studies on related species. Lessons will be learned on the critical factors in need of study and key features will enable the rapid evaluation of patterns in nature and the processes driving the patterns.

My intent for many years had been to find and study a gall-inducing insect abundant enough locally to enable rapid data collection on questions relating to plant–herbivore interactions, multitrophic level interactions among plants, herbivores and natural enemies, and the distribution, abundance, and population dynamics of species. When I moved to Flagstaff, Arizona, in 1979, I found the species that has become the focus of my studies ever since: a member of the common sawflies, Family Tenthredinidae, Order Hymenoptera. The species has no common name, but its Latin binomial is *Euura lasiolepis*: *eu* is Greek for "good" and *uro* is derived from the Greek for "tail." Hence, a "good tail" on its only host plant, the arroyo willow, *Salix lasiolepis* (Family Salicaceae). The fine "tail" is actually a very long, intricately sculptured, saw-like ovipositor used for injecting eggs into the host plant. This is a key feature of the sawfly that affects a female's behavior, its relationship to the host plant, the location of larval feeding sites, and the demography of each generation. As will be explained later, the piercing saw-like ovipositor is a phylogenetic constraint for the genus *Euura* and related genera.

ADVANTAGES IN STUDYING GALL-INDUCING HERBIVORES

Gall-inducing insects may appear to be a strange choice for deriving general principles in ecology because, to the uninitiated, they appear to be unusual. However, every group of insects provides both opportunities and difficulties for study and the advantages of gall inducers far outweigh the disadvantage of any negative bias that misguided colleagues may hold (see also Crespi *et al.* 1997). First of all, gall inducers are very speciose locally and globally; most are undescribed but an estimate of 100 000 species on earth would be modest. They are broadly distributed geographically (Price *et al.* 1998b) and can become abundant locally. Several families of insects are involved with the gall-inducing habit, as well as mites, nematodes, fungi, and bacteria. Among the insects the following are the major taxa: thrips (Thysanoptera: Phlaeothripidae), gall midges (Diptera: Cecidomyiidae), gall wasps (Hymenoptera: Cynipidae), fruit flies (Diptera: Tephritidae), common sawflies (Hymenoptera: Tenthredinidae), aphids (Aphididae), adelgids (Adelgidae), and some moth species in families such as the Gelechiidae. As a way of feeding on plants, the gall-inducing habit has evolved independently many times, providing a rich basis for comparative studies. And within many of the taxa there are many species also available for comparative work on population ecology.

The other advantages of gall-inducing insects for ecological study are multiple, and for *Euura lasiolepis* this is particularly so.

1. The female initiates gall formation before laying an egg, and the gall becomes fully formed under her influence; larval feeding is not essential for gall development. Therefore, we can evaluate exactly where a female chooses to lay eggs. We can measure her **ovipositional preference** in relation to plant quality variation: shoot length, growth rate, phenology, and chemical constituents both nutritional and defensive. We can test for preference in relation to variation in clone, genotype, ramet age, aspect, habitat, etc. (Fig. 3.1).

2. The larva lives its complete life in the gall, initially hatching and then feeding on the parenchymatous tissue of the gall. It spins a cocoon in the gall in the fall, overwinters, pupates in the spring, and emerges as an adult one year after oviposition (Price and Craig 1984). Therefore, we can readily evaluate the survival of larvae, an important component of its **larval performance** in relation to the place where the female laid the egg. We can estimate the degree to which female ovipositional preference is linked to larval performance.

3. By collecting a cohort of galls at the end of the life cycle, a complete survivorship curve and life table can be developed because

Fig. 3.1. Galls of *Euura lasiolepis* on the host plant *Salix lasiolepis*. Galls are illustrated on two clones growing side by side. Five shoots from a highly favorable male clone (NP8) are on the left and six shoots of an unfavorable female clone (NP9) are on the right. On the left many galls per shoot are large and in aggregate suppress host plant sexual reproduction. On the right only a few small galls are formed and sexual reproduction is evident as catkins. Survival of larvae in this 1979 generation was much higher on the NP8 clone (59%) than on the NP9 clone (11%). (From Price 1992a.)

time of death and cause of death can be identified by opening each gall; or survival, weight, and adult sex may be recorded. Such demographic information developed for sawflies on individual willow clones and for each generation provides critical information on variation from clone to clone, place to place, and year to year, and essential basic data on population ecology.

4. Indeed, much of the information we need for population studies of herbivorous insects is recorded in nature, if only we can learn to read the signs. Woody temperate plants, like *Salix lasiolepis*, record their annual growth over the years both in terms of annual growth rings of xylem, and in shoot length as winter bud scars mark the end of one year's growth and the start of the subsequent year. Therefore, age of ramets and growth rate per season can be estimated, and the response of female sawflies to variation in these traits can be measured. Gall growth rate, ultimate size, and contents can all be recorded, including the fate of the larva.

5. For population census work, galls are easily observed, especially during the winter when leaves have dropped. The relatively large galls of *Euura* on stems, measuring about 2 cm long and 5–10 mm in diameter, may be nondestructively, visually censused to estimate density per shoot, per ramet, per clone, and for each habitat and year. Such sampling is simple and rapid, and one sample in the winter provides a good population estimate for the generation.

6. Galls collected in the spring, before emergence begins, can be used for rearing out the contents. Sawflies, parasitoids, and inquilines can be reared, sexes can be determined for sex ratio estimates, and body mass of adults may be estimated. Thus, as well as larval survival, additional aspects of larval performance can be quantified, such as mass and gender, and the relation to host plant qualitative characters.

7. For evaluating the importance of phytochemistry in the plant–herbivore interaction, insects that form galls provide particular benefits. We can evaluate chemical constituents that may stimulate oviposition in relation to shoot length and clonal origin. For willows, the chemicals are likely to be phenolic glucosides, which are relatively well studied. And when larvae die in the category so often called "unknown causes," because plant traits of consequence are unknown and mortality caused by carnivores is not apparent, it is possible to evaluate the plant effects located in the gall that are responsible.

8. The effects on demography and population dynamics from plant to insect herbivore, or bottom-up effects, and the downward effects in the trophic system from enemies, or top-down effects, may each be evaluated effectively. We can evaluate where female sawflies will or will not lay eggs based on plant quality, and whether larvae die because of some inadequacy of the plant host module the gall is located on. Also, every impact from the top down can be quantified because predators must enter the gall, leaving a signature, or parasitoids and inquilines can be found in the gall when dissected. Winter browsing by deer and elk also leave its mark, although to sample the effects requires late fall sampling before browsing and early spring census after browsing.

Overall, gall-inducing insects are simple to study in the field and readily yield a wealth of information on the distribution, abundance, and dynamics of the species.

THE HOST PLANT

The only host plant utilized by *Euura lasiolepis* is the arroyo willow, *Salix lasiolepis*. In the Flagstaff area this willow grows as a shrub and through layering of low branches, clones spread over several square meters in favorable sites. We use Harper's (1977) terminology in defining essential components in the host plant population. A **genet** is a distinct genotype represented by one individual plant, derived from a seed, that may have spread extensively by vegetative expansion of a clone. Male and female genets coexist. A **ramet** is one main clonal component of the genet, such as a main stem or rootstock. In *Salix lasiolepis* a layered branch gives rise to a new rootstock vegetatively, and we use the term ramet to denote individual main stems emerging from these rootstocks. Each ramet passes through age classes from one year of growth to about ten years, when senescence and death of the ramet follow. Ramets can be accurately aged nondestructively by counting the number of winter bud scars up the stem from ground level to the tip. As Harper (1977) noted, plants are made up of modules that are repeated many times to produce the form and architecture of the mature individual. Ramets, shoots, leaves, and buds are all modules that provide resources for herbivores and modules may vary significantly in size, phytochemistry, and phenology, all of interest in the plant–herbivore interaction.

From this architectural development of arroyo willow we can deduce that the population of host plants provides very heterogeneous resources for the gall-inducing sawfly (Fig. 3.2). When excavated, large horizontal stems are revealed dividing into many rootstocks with each branching into many ramets. New ramets shooting from the basal rootstock grow vegetatively and rapidly in years 1 and 2 and gradually become more sexually reproductive with age. As ramets age, growth rate declines in terms of shoot vigor, but investment in sexual reproduction becomes high, especially in female clones when catkin production and pollination is followed by ovary growth, seed maturation, and release of millions of small windborne seeds. Note the contrast in Fig. 3.2 among the long vegetative new shoots at the base of the plant and the very short unbranched, but heavily reproductive, shoots in the canopy on the oldest ramets. Thus, an ovipositing female must make decisions in relation to a wide range of module vigor, phytochemistry, reproductive condition, and phenology within willow clones, and among clones in a population, and perhaps even among populations over a landscape.

The life history design of arroyo willow in the Flagstaff area is clear enough: plants produce millions of tiny windborne seeds for wide and rapid colonization of newly available mineral soil, grow rapidly in height and vegetatively over the ground to pre-empt light and space before competitors establish, and tolerate herbivores because rapid regrowth compensates for losses. In the Flagstaff area damage to willow ramets is much greater from snow breakage and flooding than from herbivores, especially *Euura lasiolepis* (Craig *et al.* 1988a) (Fig. 3.3). And competition from other plants, especially grasses, is clearly a greater threat to early survival than herbivory. Arroyo willow fits much of the "fast growth" syndrome defined by Coley *et al.* (1985).

Shoot length is a particularly valuable synoptic index of resource conditions for ovipositing and feeding sawflies. (1) Length is easily and rapidly measured. (2) The shoots usually persist after leaves have dropped, allowing measurement in the winter and spring before emergence of sawflies (some short shoots are abscised during the growing season and during or soon after leaf abscission in the fall). (3) Shoot length correlates with many other physical traits of the shoot: basal diameter, length and width of largest leaf, internode length and diameter, and length of growth in days in the season (Table 3.1). (4) Shoot length correlates with concentration of the major chemical defense compounds, phenolic glucosides (Price *et al.* 1989) (Table 3.2). (5) Mean shoot length declines with ramet age as physiological stress increases with age (Craig *et al.* 1986, 1989) (cf. Fig. 3.2). Therefore, shoot length provides an index of many traits, and even though we may not be sure of the key ingredients of a shoot that attract sawflies or that support

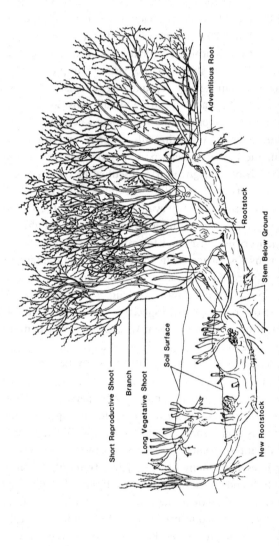

Fig. 3.2. An arroyo willow clone excavated to show the extent of vegetative propagation by layering of stems. (From Craig *et al.* 1988a.)

Table 3.1. *Traits of arroyo willow*, Salix lasiolepis, *shoots positively correlated with shoot length in 1987 (r² values are given, estimating the amount of variance accounted for by the correlation*, n = 22−25 *shoots for each clone, probabilities of no significant correlation all* <0.01)

Clone	Basal diameter	Longest leaf length	Longest leaf width	Longest internode length	Longest internode diameter	Length of growth[a]
MNA1	0.97	0.80	0.68	0.71	0.92	0.64
MNA2	0.88	0.69	0.58	0.87	0.82	0.89
CS 1	0.95	0.44	0.45	0.73	0.82	0.50
CS 2	0.89	0.61	0.45	0.72	0.82	0.68

[a] Number of days of growth.

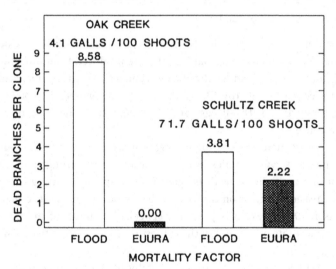

Fig. 3.3. Ramet mortality caused by flooding and snow damage (flood) and by *Euura lasiolepis* (*Euura*) in two drainages in Arizona, Oak Creek and Schultz Creek. (From Craig *et al.* 1988a.)

their larvae, we can rapidly assess this general trait and the preference among females and the performance among immatures.

THE HERBIVORE

Euura lasiolepis adults are typical of the common sawflies, with hyaline wings and a broadly linked thorax and abdomen. The smallest males weigh about 3 mg (range 3–7 mg) and the largest females a little more than 20 mg (range 7–21 mg). They are active in cool and humid

Table 3.2. *Mean concentrations of phenolic glucosides in 12 arroyo willow clones using shoot tips where sawflies attack and* r^2 *values for correlations of phenolic glucosides on mean shoot length per clone (n = 12 clones, p ≤0.01 in all cases except salicin NS)*

Phenolic glucoside	Mean concentration (mg/g dry weight)	Concentration vs. shoot length correlations r^2
Total phenolic glucosides	36.33	0.73
Fragilin	0.99	0.75
Picein	0.94	0.54
Salicin	1.60	0.03
Salicortin	15.69	0.83
Tremulacin	15.35	0.59
Tremuloidin	1.75	0.51

Source: Based on Price *et al.* 1989.

conditions, usually early in the morning and toward dusk. Sex determination is haplodiploid, that is males are produced from unfertilized eggs and females from fertilized eggs. Females are synovigenic, synthesizing eggs through the life of the adult, as opposed to proovigenic egg production with all eggs ready to oviposit at the time of adult emergence.

The female's ovipositor is long, thin, and saw-like (Fig. 3.4), as the name "sawfly" implies. An intricate design of strong tynes lined with minute, delicate teeth is beautifully adapted to cutting into soft plant tissue. But wear on the saw can be serious, reducing its efficiency (Benson 1963). Flexion is possible between the annuli of the saw, enabling accurate movement of the ovipositor through plant tissue (Smith 1968).

Females emerge in spring, generally in the morning, mate, and commence oviposition. They spend much time walking up and down leaves and along stems, antennating surfaces as they proceed. Finding the tip of a young willow shoot, they inspect it closely with both antennae and the ovipositor tip, and based on immediate cues they may decide to lay one egg through the petiole of one of the youngest leaves on a shoot and into the stem, with much undifferentiated cell tissue present, just below the meristematic tip of the shoot (Fig. 3.5). Before oviposition the female injects substances that induce gall formation, with the mechanism still to be determined.

By the time the larva emerges from the egg, the gall has grown to about half its final diameter, containing a mass of undifferentiated

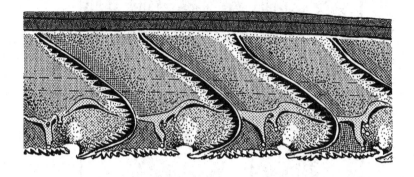

⊢————————————————————————⊣0.1mm

Fig. 3.4. The ovipositor of a *Euura lasiolepis*-type sawfly showing the saw-like shape of the long, narrow shaft, about 1.5 mm long (above) and the detailed structure of the saw (below). The egg canal is located in the dorsal section of the saw. The tip of the saw is on the left, above, and on the right, below. (From Smith 1968.)

parenchyma. The larva feeds and burrows through this mass from June through October, spinning a cocoon within the gall usually in November. Pupation occurs in May and adults emerge from galls in May and June (Price and Craig 1984). There is one more or less synchronous generation each year at any one location, with emergence coinciding with new, rapid growth of host plant shoots.

OVIPOSITION AND PREFERENCE

Euura lasiolepis females show among the highest levels of preference for host plant tissue recorded to date. First, they are found attacking only *Salix lasiolepis* in nature. Second, they oviposit only into young shoots. Third, they select the shoot modules that are growing most rapidly, which eventually become the longest shoots at the end of the season. Fourth, they prefer certain plant genotypes although these are the ones that grow most rapidly. Fifth, they prefer to lay female eggs in the highest-quality shoots. As seen in Fig. 3.2, long shoots are rare in a population of shoots on a clone, and shorter shoots are very common, but females show a strong ovipositional preference for the longest

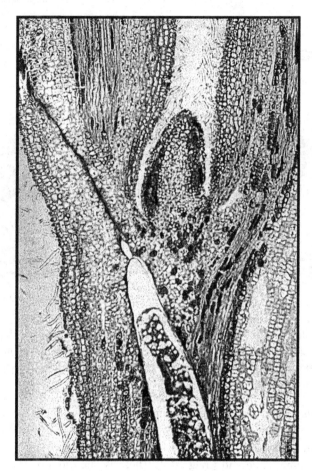

Fig. 3.5. An oviposition scar, entering at left, passes down to the egg
placed by the female just below the meristematic tip of a rapidly
growing willow stem. (Preparation and photography by Ralph Preszler.)

shoots (Fig. 3.6). With such a high preference for a low-abundance resource, the local carrying capacity for a population of sawflies is relatively low, and we should expect densities to be relatively low in willow
stands.

Females, as in most species with an arrhenotokous or haplodiploid sex ratio determination system, can make a decision to lay
a male egg or a female egg. Thus, they can influence the primary sex
ratio of progeny. In the field sex ratios varied from 39 percent males
to 60 percent males, with the higher percentages evident at the more
stressed sites (Craig *et al.* 1992). In an experiment with high and low
water treatments on willow growth, sex ratio was 40 percent male and

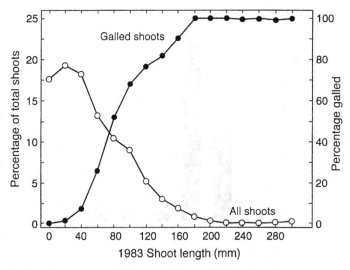

Fig. 3.6. The relationship between the availability of shoots on ten willow clones in 1983 given as the percentage of all shoots in each 20 mm shoot-length category, and the percentage of shoots galled in each category. (From Craig *et al.* 1986.)

62 percent male, respectively. On high-quality clones, which receive a higher female ratio, population growth can be more rapid and higher fitness is achieved.

In addition, more eggs are found in galls on very high-quality clones, with the probability that females choose to lay more than one egg per gall. However, we have not ruled out the possibility that more than one female contributes to the eggs in one gall.

Females select host plants and shoots using an ovipositional stimulant. This is tremulacin, a phenolic glucoside, and one of the two most concentrated glucosides in arroyo willow (Table 3.2). Tremulacin also becomes more concentrated in longer shoots compared to shorter shoots, providing a cue to females that select longer shoots (Price *et al.* 1989). As tremulacin increases in concentration, females are stimulated more to probe with the ovipositor; this phenolic is the only one that is effective in inducing oviposition (Fig. 3.7). We have not tested for additional attractants and stimulants such as CO_2 concentrations used by some herbivorous insect species to find modules with high metabolic rates and hence the most vigorous plant parts (Rasch and Rembold 1994; Stange *et al.* 1995).

Here we have a mechanistic, proximate explanation for **how** females select long shoots of *Salix lasiolepis*. We still need an ultimate

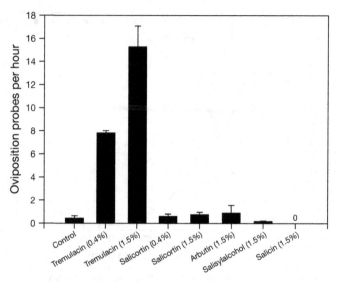

Fig. 3.7. The mean response of *Euura lasiolepis* females to various phenolic glucosides found in willows, based on the number of oviposition probes per hour. Note the strong and graded response to tremulacin at concentrations of 0.4% and 1.5%. (From Roininen *et al.* 1999.)

explanation of **why** this response should evolve, to be explained in the next section on "Larval performance."

In Chapter 7, the presence of an ovipositor in sawflies will be contrasted with its lack in the moths (Lepidoptera), so some consideration of added advantages of a plant-piercing ovipositor is worthwhile. The most important benefits are the presence of proprioreceptors and chemoreceptors which probably provide information on the internal condition of plant modules. The presence on the ovipositor of chemoreceptors enabling recognition of phenolics is based on circumstantial evidence, not morphological and physiological studies. But the circumstances strongly suggest chemical mediation of decisions by females stemming from receptors on the ovipositor. The first level of chemical reception of host plants is probably dependent on antennal receptors, as the female applies the antennae to stem and leaf surfaces as she inspects a plant. Then she flexes the ovipositor and sheath against the leaf surface as if sensing surface phytochemicals (Price and Craig 1984). This is followed by actual piercing of the leaf surface, which may result in rejection of a module or plant species, or acceptance before an egg is laid (Kolehmainen *et al.* 1994; Roininen *et al.* 1999). The fact that females reject substrates after piercing with the ovipositor implies the presence of chemoreceptors. The "saw" is "intrinsically segmented" (Fig. 3.4) with "sequentially arranged *ganglia*" and many proprioceptors, but no mention is made of chemoreceptors by Smith (1968, p. 1390). But while Chapman (1998, p. 314) notes that "the number of chemoreceptors on the ovipositor is usually very small," in insects it is hard to reconcile rejection of a particular substrate after insertion of the ovipositor with lack of an effective chemoreceptive mechanism. If the critical receptors are only the antennae and tarsi, as is generally thought to be the case (Chapman 1998), the rejection of a plant module would occur before injection of the ovipositor. Another point, suggested by Roininen *et al.* (1999) based on behavioral responses to willow species by *Euura lasiolepis*, is that ovipositional suppressants in nonhost plants may be detected by the ovipositor also. Further studies are obviously required, but these points need to be kept in mind when we consider the alternative to an ovipositor in Chapter 7.

LARVAL PERFORMANCE

This sawfly species has demonstrated one of the strongest linkages found to date between oviposition preference and larval performance (Craig *et al.* 1989). In an experiment that equalized the number of shoots per shoot length category available to ovipositing females, 65 percent of eggs were laid on the longest 20 percent of shoots and none were laid on the shortest shoot length category. In the field, survival based on 4181 galls on 15 willow clones declined rapidly from 85 percent on 1-year-old ramets to 6 percent on 9-year-old ramets, the oldest living ramets in these clones. There is a very strong negative correlation between ramet age and shoot length developed in a single season. Clearly, females prefer to oviposit on shoots in which larvae will survive the best, an ultimate, evolutionary explanation for this female behavior.

In fact, females make more behavioral decisions when engaged with the host plant than we anticipated. In experimental arenas we allowed two females per plant with about 100 eggs between them to relate to three treatments, some plants with a high-water treatment with rapid growth, some with a medium-water treatment, and some with a low-water treatment with poor growth (Preszler and Price 1988). The maternal responses to these three categories of host plant quality were dramatically different. With 100 eggs in the abdomens of females, 100 galls were formed on high-quality clones, but only 38 galls appeared on low-quality clones. In addition, careful dissection of a subsample of galls, soon after oviposition was completed, revealed that egg retention

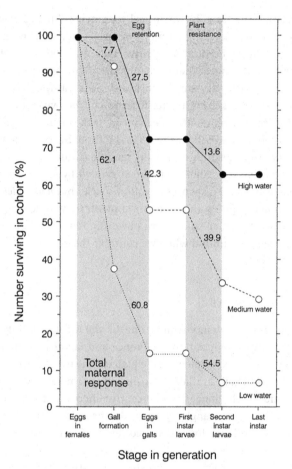

Fig. 3.8. The response of female *Euura lasiolepis* and larval survival to willow clones grown at three water treatments: high, medium, and low. Note the strong maternal response (shaded on left), an aggregate of decisions to form a gall and to lay an egg in a gall. The effects of plant resistance, or death of larvae due to deficiencies in the plant host are shaded on right. (Based on data in Preszler and Price 1988; from Price, P. W., T. P. Craig, and M. D. Hunter (1998) Population ecology of a gall-inducing sawfly, *Euura lasiolepis*, and relatives, pp. 323–340, in J. P. Dempster and I. F. G. McLean (eds.) *Insect populations: In theory and practice*, with kind permission of Kluwer Academic Publishers.)

occurred in all treatments, but was relatively low at 27.5 percent of eggs in the high-water treatment but as high as 60.8 percent of eggs in the galls initiated on low-water willows (Fig. 3.8). As a result the full maternal response to host plant quality ranged from 72 eggs in high-water hosts to 14 eggs in low-quality plants, a more than fivefold difference in egg numbers. Clearly, the evolved behaviors of females in relation to host plant quality play a major role in the abundance of galls and larvae. No other effect on abundance is likely to be stronger than host plant quality variation coupled with the evolved maternal response to this variation.

Coupled with the maternal behavior is the larval survival. On high-water treatment plants only 13.6 percent of first instar larvae died as a result of inadequate plant resources. On low-quality plants 54.5 percent of first instar larvae died soon after emergence from the egg (Fig. 3.8). This amounts to a fourfold difference in survival of larvae resulting from some form of plant resistance, although the impact on numbers per host plant is relatively small compared to the total maternal response.

Why do larvae die in galls when the only factor in the experiment is the host plant? What is the proximate explanation for death? This question is usually not addressed in the plant–herbivore literature unless toxic phytochemicals are suspected. Death is often ascribed to "unknown causes." And phenolics and tannins in woody plants are generally thought to work in a quantitative way, reducing larval performance without actually killing larvae rapidly. Therefore, the phenolics in willow were not regarded as critical in larval death, particularly when we found that galls on wild plants contained significantly lower concentrations of phenolics than ungalled tissue and higher protein concentrations (Waring and Price 1988).

We reasoned that if larvae die so rapidly after eclosion from the egg, parenchyma cells in galls on stressed plants must be unsuitable as food in some manner. A well-known response of plants to stress is to increase cellular concentrations, increasing osmotic potential. We therefore hypothesized that the osmotic potential in galls on stressed plants would become elevated above that of the *Euura* larvae such that there would be a net loss of water from larvae, and nutrients would not diffuse across the gut wall. Preliminary results are consistent with this hypothesis, but more experimental work is needed. The osmolarity of female hemolymph in which eggs form is 613 mOsm/l, and on young ramets, osmolarity in galls with dead larvae was consistently higher than that in females, while it was lower in galls with living larvae (Fig. 3.9). On old ramets, the osmolarity was more variable but one general pattern was that galls with living larvae always had a lower osmolarity than galls with larvae that died very soon after eclosion.

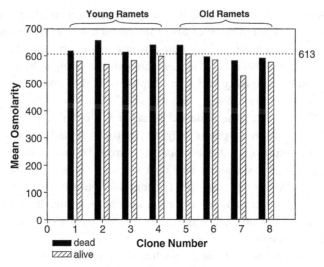

Fig. 3.9. A test of the hypothesis that osmolarity in galls defines the survival or death of newly hatched larvae of *Euura lasiolepis*. The osmolarity of female hemolymph is shown at 613 mOsm/l and the osmolarity of gall parenchyma is illustrated for eight clones, with means for galls with dead and living larvae separated. Note that on young ramets dead larvae were found when osmolarity exceeded 613 mOsm/l and in all clones osmolarity was higher when larvae died than when larvae remained alive. (K. J. Leyva, unpublished data.)

Another contributing factor to larval death, and hence selection on females that promotes oviposition into long shoots, is host plant shoot abscission (Craig *et al.* 1989). Short shoots are abscised much more frequently than long shoots (over 80 percent to 5 percent probability respectively) with a clearly linear decline in probability ($r^2 = 81.3$ percent, $p < 0.005$). When abscised shoots fall to the ground galls become saturated with water and all larvae die (Craig *et al.* 1989).

In sum, we have proximate and ultimate explanations for female preference for rapidly growing, vigorous shoots, and proximate and ultimate explanations for why there exists a strong preference–performance linkage between female oviposition and larval survival. However, the proximate explanation for larval death is only tentative.

THE PHYLOGENETIC CONSTRAINT

Female behavior in relation to the host plant is so central to the understanding of the general biology of the species that we should anticipate that a phylogenetic constraint would be evident relating to oviposition. Indeed, the ovipositor itself acts as a constraint. Its architecture is complex, efficient, but delicate, limiting oviposition to tender plant tissue. The tissue selected is the youngest that a female can find. This further limits the evolution of alternative life cycles because females must emerge when the phenology of the host plant is appropriate – early in the growth of shoots before differentiation of cells is proceeding rapidly. Another limitation is that eggs are laid inside plant tissue, accompanied by wear on the saw.

We also see that the ovipositor is a plesiomorphic character common to all tenthredinid sawflies, and all other sawfly groups with eight other families present (Gauld and Bolton 1988; Smith 1993). It is a character basic to the whole Suborder Symphyta. The ovipositor also appears to be constrained, being retained in all tenthredinid sawflies even though it becomes short in some species (Price and Carr 2000). In fact, the ovipositor design dates back to some of the earliest wingless insects, the silverfish from Devonian times (more than 350 million years ago). This lepismatid ovipositor is regarded as a groundplan trait for the Hymenoptera not found in any other insects with complete metamorphosis (Gauld and Bolton 1988; Hunt 1999).

These five characteristics, complex architecture, need for tender plant tissue, phenological synchronization with host plant, endophytic oviposition, and a plesiomorphic trait, were used by Price and Carr (2000) to argue that the shape and function of the ovipositor is the phylogenetic constraint for this and related species. It limits the evolutionary options strictly to how eggs are laid (endophytically), where the eggs are laid (on young shoots), when the eggs are laid (early in plant shoot phenology), and how frequently eggs can be laid in the season (at one phenological stage of the host), resulting in a univoltine life cycle.

Ricklefs and Miles (1994, p. 29) employ the phrase "the behavioral filter between ecology and morphology" advisedly. For the ovipositor influences many behavioral traits of the sawfly, treated next under "The adaptive syndrome," that influence directly emergent properties explained in Chapter 4.

THE ADAPTIVE SYNDROME

The set of coordinated derived characters that cluster around the ovipositor and oviposition as a phylogenetic constraint is extensive. They mitigate the ecological limits imposed by the ovipositor. We emphasize life history traits and behaviors and list them as in Price and Carr (2000), with an additional seventh point added here.

1. Young and soft plant tissues are selected as oviposition sites, minimizing wear on the saw.

2. The life history of univoltine sawfly species is adapted to host plant phenology, which maximizes resource availability of young shoots that are growing rapidly.

3. Sensory receptors in the females ... can detect long vigorous shoots and the specific host plant species (Roininen et al. 1999).

4. Oviposition preferences in the females relating to shoot length and vigor ... maximize larval performance. There is a strong link between preference and performance.

5. Being haplodiploid, a plesiomorphic character for the Order Hymenoptera, females can allocate the sex ratio of progeny according to the quality of shoots available. Such allocation has become part of the adaptive syndrome, not of course the plesiomorphic sex-determination system itself.

6. Oviposition into plant tissue predisposes a lineage to become gall inducers, and living in a gall has several adaptive features (Price et al. 1987a).

7. With a female selecting shoots very carefully and laying eggs endophytically while determining the sex of each egg, the most effective allocation of a limited number of eggs is to place them individually. Long searching times and a limited period each day allows egg production to be relatively low, and production is synovigenic. About one egg per ovariole, with 12 ovarioles per female, is matured each day, and a female may live for only four or five days (Price and Craig 1984; Craig et al. 1992). A fecundity of about 50 eggs per female per lifetime is a good general estimate, if she dies of old age.

The flow of effects from the phylogenetic constraint to the adaptive syndrome in *Euura lasiolepis* can best be illustrated in a figure (Fig. 3.10). Clearly, these evolved characters have a strong effect on the ecology of the species, the emergent properties. These are treated in the next chapter.

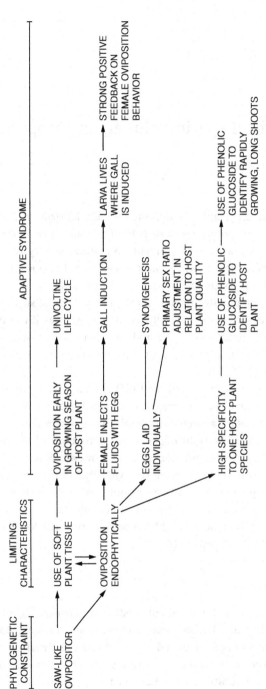

Fig. 3.10. The hypothesized web of influences from the phylogenetic constraint of a saw-like ovipositor in *Euura lasiolepis* to the coordinated adaptations representing the adaptive syndrome for this species.

4

The focal species – Emergent properties

Based on the adaptive syndrome of *Euura lasiolepis*, it is evident that the emergent properties of distribution, abundance, and population dynamics are all very dependent on the availability of suitable resources provided by the willow host population. These resources are highly variable in space and time, the critical factor being the local production of the rapidly growing, more juvenile type of shoots. These high-quality resources are generally in short supply, as illustrated for well-established clones in Figs. 3.2 and 3.6. Therefore, an appreciation of willow module variation over a landscape is necessary to understand the ecology of the sawfly.

In Chapter 3 I emphasized the link between female ovipositional preference and larval performance. Understanding this close relationship was important for a mechanistic explanation of why females would be so selective, thereby resulting in severe limitation of resources to individuals and to the population at large. However, only female preference is critical to understanding the typically low carrying capacity in local willow populations and how resource supply can change in response to water availability to the willow host plants. Of course, when a female's choice is compromised by poor resources, the local result is observed as reduced larval survival.

RESOURCE VARIATION

Three main factors influence shoot module production: disturbance, water availability, and willow ramet and genet age. Erosion, flooding, fire, and snow damage all played their role in the dynamics of the willow populations. Any disturbance resulting in the exposure of mineral soil created a substrate for willow seed germination, provided the site remained wet enough. In the Flagstaff area seedlings needed three good years of summer and winter precipitation for establishment, a rare event occurring perhaps only once every 100 years (Sacchi and Price 1992). The establishment of young populations of willow with high-quality resources for sawflies would have been infrequent and patchy. Fire also undoubtedly played its role in the module demography of arroyo willow, burning into willow thickets from adjacent ponderosa pine stands, which burned every three to five years in the vicinity before human settlement (cf. Stein et al. 1992). Resprouting of willow after fire produces very favorable shoots for herbivores, be they grasshoppers, sawflies, or deer and elk. Since human settlement such disturbances have been greatly reduced by controlling water courses and fires. And although human disturbances of other kinds have impacted landscapes in catastrophic ways, the general effect is to reduce good habitat for willow and to reduce the creation of new habitat.

Arroyo willow is limited to riparian habitats, springs and areas that collect water such as borrow pits (e.g. pits resulting from removal of soil for construction of roads) and natural depressions. Water availability in the Flagstaff area varies significantly from permanently running creeks and springs to ephemeral drainages with running water, perhaps for six weeks in some years. As a result willow growth differs based on the habitat it occurs in, with better growth in wetter sites. Added to this habitat variation is the temporal variation due to changes in winter precipitation. High winter precipitation, measured from October to May, results in better willow growth in the subsequent season. Better willow growth includes shoots growing more vigorously and longer, more shoots are initiated, and more new juvenile ramets are initiated, all contributing to a marked increase in resource quantity and quality. These effects of water on willow growth have been documented in the field and in experiments (e.g. Price and Clancy 1986a; Preszler and Price 1988; Waring and Price 1988; Craig et al. 1989).

In general, high water supply in the soil improves willow growth and resources for Euura lasiolepis. Populations are higher in wet sites than in dry sites and after high winter precipitation. The chain of effects can be illustrated after high winter precipitation, as an example (Fig. 4.1). After a winter with high snowfall, during sawfly generation t in the Flagstaff area, willows grow better, having multiple positive effects on the second trophic level of sawflies, all contributing to an increase in the next generation of sawflies in generation $t + 1$. In addition, in generation $t + 1$, the higher survival of larvae in galls, the higher proportion of females in galls, and females of a larger size all are likely to contribute to higher populations in generation $t + 2$. Such effects

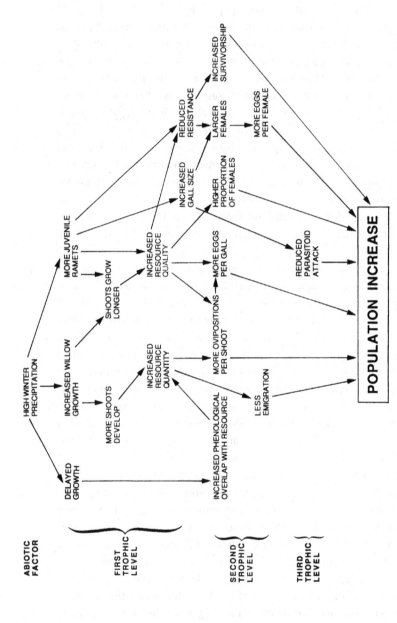

Fig. 4.1. The flow of effects after high winter precipitation on arroyo willow growth, *Euura lasiolepis* characteristics, and natural enemies, all increasing the probability of a population increase. (From Price 1992a.)

were determined mechanistically by Price and Clancy (1986a) and the delayed effects appear in the time-series analysis treated later in this chapter.

These very strong bottom-up effects through the trophic system dictate the population dynamics of the sawfly. Annual censuses of galls on 15 individual clones over 19 years from 1981 to 1999 show that populations differ dramatically between clones in wet and dry sites for most of the period and change significantly through time (Fig. 4.2). Even within particular sites clones differ in general population densities (Fig. 4.3). In all 15 clones change in populations is more or less synchronous, driven by winter precipitation (Figs. 4.2–4.5).

The effects of low precipitation in the winters of 1984, 1989, and 1996 can be observed on all clones. Clones near permanent springs (CS1 and CS2, CS = Coyote Spring), and BD1 which receives runoff from the Biology Department roof at the Museum of Northern Arizona all had persistently higher populations until 1995 than any of the other clones, which were located along Schultz Creek, a temporary stream (cf. Fig. 4.2, BD1, CS1, CS2 compared to NP7 and MNA6 and the clones in Figs. 4.3–4.5; see also map in Fig. 4.6). Among clones growing side by side differences tend to persist over many years, best shown in Fig. 4.3. MNA1, 2, and 3 grew touching each other for most of the study period but MNA2 consistently supported higher populations than MNA1 and 3 in any one year. MNA2 is a male clone, resulting in rapid flowering, dehiscence of catkins, and relatively early vegetative growth with better phenology for sawfly attack. MNA1 and 3 are female clones, which flower and mature seed, with delayed vegetative growth relative to males and reduced suitability for *Euura* oviposition. In addition senescence and death of ramets of MNA2 occurred earlier than in the female clones, with stronger recruitment of new ramets, keeping the male in a more vigorous state of vegetative growth than the females.

Over all 15 clones sampled, there existed a remarkable persistence of relative abundance of galls for the 11 years tested (Price *et al.* 1995a) (Table 4.1). Correlations of abundance in each clone from one year to another and across all clones per year showed high consistency for 11 generations. Even comparisons of gall densities 11 generations apart (e.g. 1983 versus 1993) showed remarkably high predictability of densities in one year by densities in another, with 99 percent of the variance accounted for in the 1983/93 comparison. Even when relatively low precipitation occurred in the winters of 1987–9, significant correlations persisted except in two cases (r^2 values of 0.21 and 0.25 within the dashed line box). For the clones growing well at present such differences

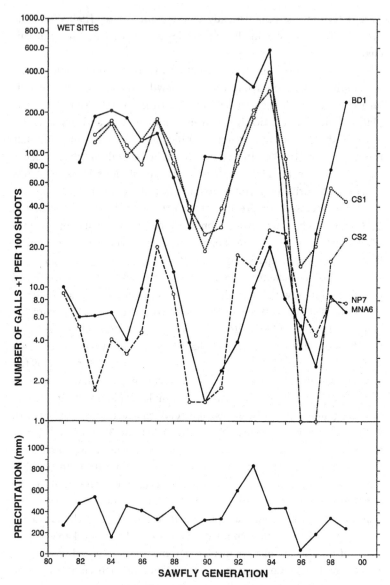

Fig. 4.2. Populations of *Euura lasiolepis* on individual willow clones. Clones BD1, CS1, and CS2 occurred in wetter sites and NP7 and MNA6 were located along the drier Schultz Creek drainage in Flagstaff, Arizona. Precipitation from October to May during each sawfly generation is recorded below for the Flagstaff area.

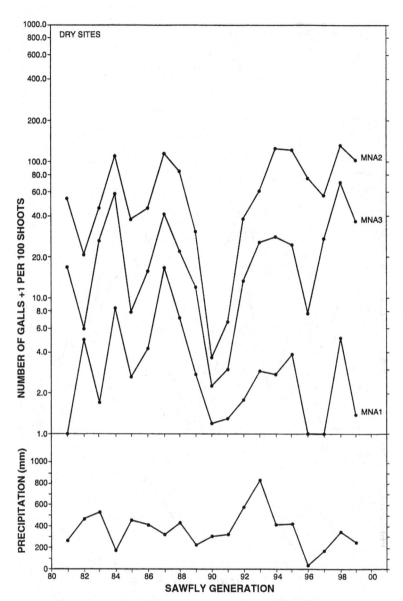

Fig. 4.3. Populations of *Euura lasiolepis* on willow clones MNA1, MNA2, and MNA3 growing adjacent to each other along Schultz Creek.

persist into 2002, indicating approximately the same ranking in densities over 15 clones for 22 years. Such consistency clearly results from a combination of common responses of the willow and sawfly populations to (1) variation in winter precipitation, (2) the strong effect of

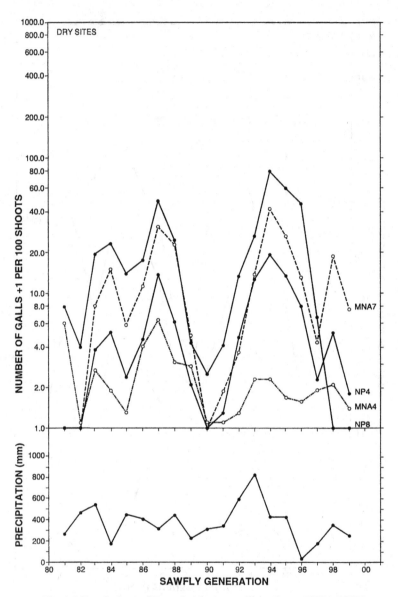

Fig. 4.4. Populations of *Euura lasiolepis* on willow clones MNA4, MNA7, NP4, and NP8, all located along Schultz Creek.

habitat variation between wet and dry sites, and (3) individual clonal differences.

In addition to the natural effects of winter precipitation, habitat and clonal variation, there were anthropogenic effects. These included introduction of an exotic pest, replacement of a water pipeline beside

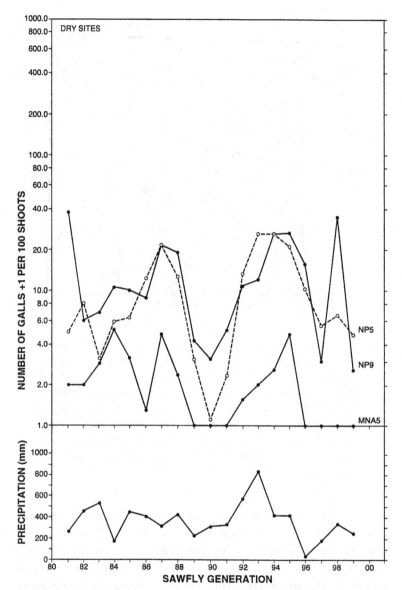

Fig. 4.5. Populations of *Euura lasiolepis* on willow clones MNA5, NP5, and NP9. These clones occurred in drier sites along Schultz Creek.

Schultz Creek, and cutting back willows along the roadside. The oystershell scale, *Lepidosaphes ulmi*, is a serious pest of woody angiosperms introduced from Europe at an unknown time (Mattson *et al.* 1994). It has been in the Flagstaff area for decades but has spread among the

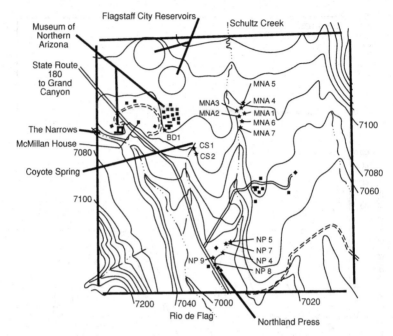

Fig. 4.6. Map of clone distribution in the northern part of Flagstaff, Arizona. MNA signified Museum of Northern Arizona clones, BD1 is the clone near the former Biology Department at the Museum, CS1 and CS2 are clones growing at Coyote Spring, a permanently running source of water. NP clones occur in the vicinity of Northland Press. The McMillan House Spring was the site used for studies on the window of vulnerability of *Euura* larvae to parasitoids (see Fig. 4.9). The map covers a section of land 1 mile on the side, or 1630 m; the numbered lines are altitude contours (in feet).

arroyo willow population rapidly only in the past 15 years. The scale becomes so dense on stems that ramets are killed, but presence of the scale is limited largely to wet sites. I first noticed the scale on clone BD1 in 1979, and since then it has spread to CS1 and 2. Death of ramets on these clones became so prevalent around the time of the 1996 drought that populations of sawflies plummeted (Fig. 4.2). None were censused on CS2 in 1996 and 1997, when shoots available for attack were absent. The scale also became less dense on new ramets after 1996 and willows recovered gradually, producing young, vigorous ramets very suitable for sawfly attack, so populations have rebounded in 1998 and 1999. A water pipeline was replaced along Schultz Creek during the 1995 sawfly generation. Clone MNA5 was removed entirely and never resprouted (Fig. 4.5).

Table 4.1. *Correlation matrix[a] among 11 generations of* Euura lasiolepis *(1983–1993) on 15 clones of* Salix lasiolepis *in the Flagstaff, Arizona area*

Year	1984	1985	1986	1987	1988	1989	1990	1991	1992	1993
1983	0.91	0.99	0.89	0.75	0.55	0.66	0.85	0.91	0.87	0.99
1984		0.90	0.93	0.94	0.80	0.89	0.61	0.70	0.64	0.91
1985			0.89	0.77	0.56	0.66	0.84	0.91	0.95	0.99
1986				0.89	0.76	0.79	0.60	0.71	0.63	0.91
1987					0.93	0.96	0.40	0.50	0.44	0.78
1988						0.94	0.21	0.30	0.25	0.57
1989							0.30	0.38	0.33	0.67
1990								0.98	0.99	0.82
1991									0.98	0.89
1992										0.84

[a] Values in the table are correlation coefficients squared, r^2, providing an estimate of the proportion of the variance accounted for by the linear correlation between two years. Critical values for r^2 are 0.264, $p < 0.05$; and 0.411, $p < 0.01$. The dashed line encompasses values where 50% or less of the variance is accounted for, during a series of drought years, 1987–9.

Source: Price *et al.* (1995a).

MNA1 had all above-ground growth removed, so the sawfly population went to zero; the clone resprouted from rootstocks and shoots were colonized by the 1998 generation. Two other clones were depleted in size, MNA3 and 4, but impact on sawfly densities was not evident. Another human impact occurred on NP8 and 9, adjacent to State Route 180 when the clones were cut back in 1996 and NP8 was cut to ground level completely in 1997 and 1998, causing the death of most of the clone and the demise of the *Euura* population in that clone (Fig. 4.4). NP9 was less impacted because parts of the clone are further from the road and remained uncut (Fig. 4.5), but in 2001 NP9 was also demolished during road widening.

Heterogeneity of resources for sawflies is evident within clones, among clones at the same site growing adjacently, among sites depending on water availability, from year to year based on winter precipitation, and because of introduced pests and human management practices. It is these bottom-up forces from resources to the sawfly population that dictate the distribution, abundance, and population dynamics of *Euura lasiolepis*. Bottom-up effects are very strong in this trophic system. What role does this leave for carnivores to play as top-down effects of the natural enemies?

NATURAL ENEMIES

The impact of natural enemies is very weak. Carnivores may change the amplitude of fluctuations slightly but they do not influence the fundamental dynamics. Several factors are involved here, the most important being the very strong influences of clonal, ramet, and shoot quality variation. A consequence of this is that populations on vigorous clones tend to induce larger galls, which offer more protection against small parasitic wasps, the parasitoids, such as pteromalids (Price and Clancy 1986b). For larger parasitoids, such as ichneumonids, galls toughen rapidly, reducing the window of vulnerability of the larval inhabitants (Craig et al. 1990a). Vertebrate predators, such as mountain chickadees, may cause high mortality but very locally and sporadically. Winter browsing by elk and deer involves the consumption of some galls, but again such browsing is very patchy and sporadic, being concentrated in dry winters.

Because galls become larger on very vigorous shoots, such shoots are most common in wet sites or after heavy winter snows, and sawfly populations in the sites are relatively high, there is a negative correlation between sawfly density and attack by small parasitoids (Price and Clancy 1986b; Price 1988) (Fig. 4.7). Large galls formed on vigorous shoots provide a refuge from attack at about 6.5 mm diameter when observed attacks begin to fall well below expected frequencies based on the hypothesis of random attack and the frequency of gall diameter classes in the population. Thus, the correlation between gall diameter class and percentage parasitism is negative, significant, and evident in most years.

For larger parasitoids such as the ichneumonid *Lathrostizus euurae*, gall diameter does not provide a refuge to *Euura* larvae, but toughness does (Craig et al. 1990a). This parasitoid has a long and flexible ovipositor, inserted and drilled into the gall as the female rotates (Fig. 4.8). And as galls toughen with age, access to the parasitoid decreases (Fig. 4.9). The "window of vulnerability" of *Euura* larvae is therefore defined by the date of hatching from the egg to the time when galls are too tough to drill, which is correlated with gall diameter. The correlation between toughness and diameter differs among clones, making the window of vulnerability variable among clones: the larger the window, the higher the parasitism (Craig et al. 1990a).

The net result of the diameter effect on small parasitic wasps and the toughness effect on larger parasitoids minimizes their impact, such that significant top-down effects are undetectable in experimental

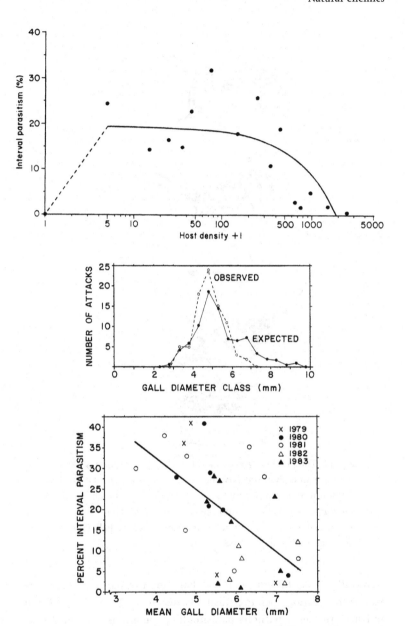

Fig. 4.7. Relationships among host density and gall diameter classes and the number of attacks by parasitoids. Interval parasitism shows the percentage of mid-sized larvae parasitized as this is the stage at which attack occurs. (Top is from Price, P. W. (1988) Inversely density-dependent parasitism: The role of plant refuges for hosts, Fig. 2, *J. Anim. Ecol.* 57: 89–96, Blackwell Science, Oxford; middle and bottom are from Price and Clancy 1986b.)

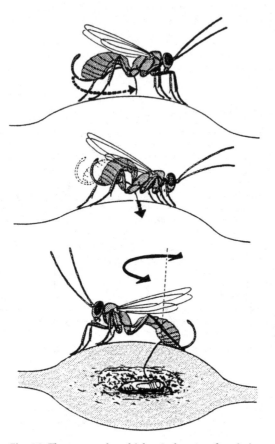

Fig. 4.8. The process by which a *Lathrostizus* female inserts her ovipositor to locate an early instar larva of *Euura* in a gall. First, she unsheaths her ovipositor and brings it into position between the two hind legs. Then she forces the ovipositor into the gall using movements of the abdomen, and finally she twists around to fully penetrate the gall. (Slightly modified from Kopelke 1988.)

arenas (Fig. 4.10). In willow clones that were divided into three parts, each with a section open to parasitoids, another caged with parasitoids, and a third caged without parasitoids, populations remained statistically similar for three generations (Woodman 1990). We could not reject the hypothesis that top-down impact by parasitoids is negligible. Although parasitoid species richness per host insect and impact on host mortality are both intermediate relative to other feeding types (e.g. free-feeders and leaf miners), observational and experimental evidence indicates weak effects in this system (cf. Hawkins and Lawton 1987; Hawkins 1988, 1994).

Other effects of predators by chickadees pecking open galls and extracting larvae, ants eating sawfly adults, and grasshoppers, elk, and deer eating galls are inconsequential or have not been estimated adequately (cf. Woodman and Price 1992; Hunter and Price 1998).

LATERAL EFFECTS

Lateral effects in trophic systems, from conspecific or heterospecific individuals, require some care with attributing the direction of influence. Many lateral effects are actually attributable to shortage of food resources, and a bottom-up influence should be assigned. If effects occur at densities well below that where food supply becomes limiting, then a purely lateral effect is probably operational. Such effects would include, perhaps, interference competition, cannibalism, territoriality, and dispersal (Harrison and Cappuccino 1995). These possibilities will be treated in turn in relation to *Euura lasiolepis*, but cannibalism has not been observed.

Intraspecific competition may occur among females for sites to insert eggs, or indirectly among larvae when galls become numerous enough on a stem to deplete resources for the most distal galls. Females compete at all densities of sawflies because the longest shoots are preferred (Craig *et al.* 1990b). Once the best sites are occupied by eggs and oviposition scars, lower-quality shoots are utilized, and, as sawfly densities rise, so fewer eggs are laid per female. The mechanism of competition is an unusual form of territoriality in which plant wound compounds from the oviposition scar become repellent to subsequent females (Craig *et al.* 1988b). The scar darkens slightly and after two hours from oviposition the scar inhibits further ovipositions, suggesting that phenolic glucosides are oxidizing to highly toxic quinones. This form of interference competition through a kind of territoriality may well be regarded as a lateral effect, although because the best-quality shoots are invariably limiting in current landscapes, there is a strong bottom-up element at work. Whether such competition actually influences distribution, abundance, and population dynamics in important ways is debatable. Obviously, competition will be higher where populations are already high and this negative feedback will be density dependent. But ascribing an important role to competition when high densities are reached only where resources are very favorable, argues

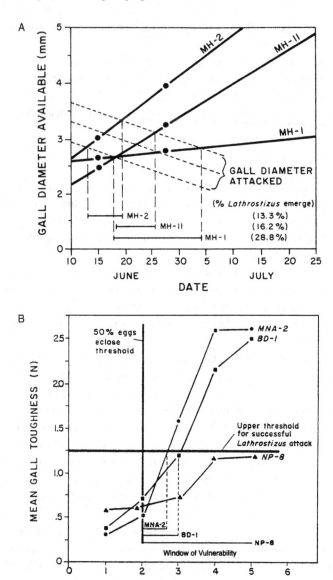

Fig. 4.9. The window of vulnerability for *Euura* larvae in relation to parasitoid attack by the ichneumonid *Lathrostizus euurae*, from studies by Craig *et al.* (1990a) and Craig (1994). (A) Dates are when *Euura* larvae are in galls and galls are expanding in diameter and toughening. Hence, as days pass the gall diameters attacked actually decrease as galls toughen. Therefore, galls that increased in diameter and toughened rapidly, represented by clone MH2, produced a short window of vulnerability and relatively low parasitism by *Lathrostizus* (13.3 percent). Galls that grew

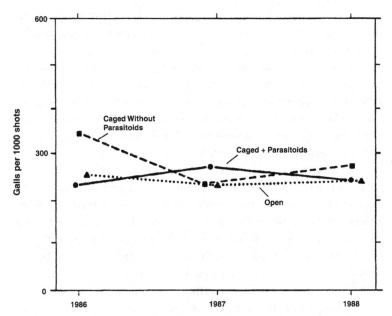

Fig. 4.10. Densities of sawflies for three generations on clones divided into three parts: open to emigration and immigration of sawflies and parasitoids, caged without parasitoids, and caged with the full complement of parasitoids. (Based on Woodman 1990, from Price 1990a, b.)

that resource limitation, a bottom-up influence, is really the key to understanding this interaction.

A similar situation prevails when indirect larval competition is considered. In one study significant effects of gall density on larval survival were not detected (Craig *et al.* 1990b) and in another effects

Caption for fig. 4.9 (*cont.*)

and toughened slowly, as in clone MH1, provided a wider window of vulnerability and resulted in higher rates of parasitism (28.8 percent). (B) Model for estimating the window of vulnerability of galls on three willow clones (MNA2, BD1, and NP8). The horizontal line represents the toughest gall in which a *Lathrostizus* egg was found in an *Euura lasiolepis* larva. Toughness was measured by a penetrometer in newtons (N), a unit of force that produces an acceleration of 1 meter per second on a mass of 1 kilogram. The vertical line represents the week in which 50 percent of galls contained larvae. The week in which 50 percent of galls had larvae did not vary among clones, but the rate of increase in toughness did. Note the different window lengths shown at the bottom, which resulted from differing rates of increase in toughness.

were observed only after the sixth gall on a shoot. Over the years and clones, more than six galls per shoot is infrequent enough to affect population dynamics little.

Euura lasiolepis has a negative effect on other gall-inducing sawflies in some situations (Fritz *et al.* 1986) but not in others (Fritz and Price 1990). Other galling species apparently had no effect on *Euura* partly because *Euura* was the first species to oviposit, and because it occupies basal stem positions and all other sawflies are distal and peripheral on a shoot (see next chapter for details).

Considering dispersal, females emerge from galls in spring and fly about in the canopy of the natal clone, searching for oviposition sites after mating. They are not commonly seen moving between clones, although when experimental plants were small, dispersal by females commonly occurred up to 8 m (Stein *et al.* 1994). Beyond this distance a small percentage of females was found. Female *Euura* appear to be very philopatric, focusing movement on the natal clone, although clones with high sawfly densities generally act as sources for nearby clones with low densities. This source–sink relationship at a very local level tends to stabilize populations as "sharing" of sawflies among clones reduces the rates of increase and decline of populations.

Overall, lateral effects as emergent properties contributing to population dynamics appear to be weak in this system except for those discussed in the next section.

POSITIVE FEEDBACK LOOPS

As Hunter (1992b) has emphasized, followed by Hunter and Price (1992a), feedback loops from herbivores to plants often play a role in creating greater resource heterogeneity for consumers. This is certainly the case in *Euura* populations, with two phenomena of note. One we called **resource regulation** by sawflies and the other **facilitation**. In both cases the gall inducers change plant quality very locally, improving resources for subsequent generations or for siblings perhaps.

Resource regulation occurs when gall densities increase to relatively high levels (Craig *et al.* 1986). When adults emerge from galls, leaving a large exit hole, fungi invade and necrosis sets in. Much tissue damage results, not confined to the gall tissue, for vascular tissues become invaded and killed, resulting in death of the distal parts of ramets. Such infections may kill single shoots, but more often death is delayed and ramet tips decline in vigor over two or more years, resulting in an effective pruning of several years of distal growth. Regrowth then occurs more proximally to the rootstock and is more juvenile than in the normal aging of a ramet. Hence, the current local population benefits from this form of resource regulation, which acts as a positive, density-dependent feedback loop.

Facilitation, the second form of positive, density-dependent feedback loop (Craig *et al.* 1990b), functions in the short term within generations and may well promote the fitness of siblings or even the same female that initiates the facilitation process. When a gall is induced, the mechanisms of which are yet to be explained, local cell division and growth are stimulated. The galled internode increases in length beyond its normal growth, but also the next more distal internode is stimulated. This induced vigor then is preferred by subsequent females, or the original female if she returns to the site within a day or two. The net result is that galled willow plants receive more ovipositions than ungalled plants when plants are of the same quality and the treatment is simply galled or ungalled plants at the initiation of the experiment. The effect is also clearly visible in natural populations where increased internode elongation distal to a new gall creates a highly preferred site for oviposition.

These positive feedback loops work more effectively as populations increase. At high populations resource regulation can maintain high resource quality more or less indefinitely. The positive feedback of resource regulation then becomes balanced by the negative feedback of competition as far as that is permitted by the highly variable effects of water availability to host plants.

A summary of experiments, results, and sources supporting the scenarios described in Chapters 3 and 4 is provided in Table 4.2.

ANALYSIS OF POPULATION DYNAMICS

The above catalog of interactions indicates very strong bottom-up plant resource effects from water availability, plant age, and disturbance. All effects relate fundamentally to the water and nutrients available for growth to each genet, ramet, and smaller modules. Therefore, in a time-series analysis we should expect a strong general effect of winter precipitation, with time delays included. Time delays would result from winter precipitation's effect on the next generation of sawflies and on the next season's willow growth and the one following perhaps.

Table 4.2. *Experimental studies on the sawfly* Euura lasiolepis *and its host plant,* Salix lasiolepis

Type	Result	Reference
Bottom-up effects		
Water treatments on plants	High water: higher gall densities and higher survival	Price and Clancy 1986a
	Preference–performance linkage	Craig *et al.* 1989
	High water: more galls, more eggs in galls, and higher survival	Preszler and Price 1988; Price 1990a
Fertilizer and water treatments	Shoot growth best predictor of sawfly preference and performance	Waring and Price 1988
Clonal phenotypic effects of interspecific competition	All species increase as shoot length increases	Fritz and Price 1990; Fritz *et al.* 1986
Oviposition stimulant	Phenolic glucoside, tremulacin, only effective stimulant	Roininen *et al.* 1999
Oviposition deterrent	Oviposition scars become deterrent to subsequent females	Craig *et al.* 1988b
Pruning to stimulate browsing herbivory	Shoot length and gall density increase	Hjältén and Price 1996, 1997
Fire	Rapid regrowth after fire, highly palatable	Stein *et al.* 1992
Preference–performance	Strongest linkage recorded to date in literature	Craig *et al.* 1989
Population perturbation	Rapid decline to background levels	Price *et al.* 1995b
Sex ratio control	Female biased ratios in high-quality modules	Craig *et al.* 1992
Lateral effects		
Competition for oviposition sites	Oviposition scars become repellent to subsequent females	Craig *et al.* 1988b, 1990b
Facilitation	Gall initiation facilitates subsequent shoot growth	Craig *et al.* 1990b
Dispersal	Females highly philopatric to natal site	Stein *et al.* 1994

Table 4.2. (*cont.*)

Type	Result	Reference
Top-down effects		
Parasitoid exclusion	No change in host density in three generations	Woodman 1990; Price 1990a
Window of vulnerability to parasitoid	Narrow window imposed by gall toughening	Craig *et al.* 1990a
Gall size and parasitoid attack	Larger galls reduce probability of attack	Price and Clancy 1986b

Source: Slightly modified from Price *et al.* (1998a). Reproduced from Price, P. W., T. P. Craig, and M. D. Hunter (1998) Population ecology of a gall-inducing sawfly, *Euura lasiolepis*, and relatives, pp. 323–340, in J. P. Dempster and I. F. G. McLean (eds.) *Insect populations: In theory and practice*, with kind permission of Kluwer Academic Publishers.

These expectations are fully justified by the analysis covering the first 16 years of study on *Euura lasiolepis* while sampling methods were standardized (Hunter and Price 1998). Fluctuations in winter precipitation are clearly followed, with a delay, by sawfly population fluctuations (Fig. 4.11). An apparent cycling of winter precipitation results in an apparently cyclic behavior in sawfly populations. When sawfly density in time $t + 1$ is plotted against the independent variable of winter precipitation during generation t, there is a clear correlation in both dry-site clones (MNA1–7, NP4, 5, 7–9) and wet-site clones (BD1, CS1 and 2) (Fig. 4.12). In addition, when sawfly density in $t + 2$ is correlated with precipitation in time t, there is additional variation accounted for in dry-site clones, and the regression is positive but not significant in wet-site clones (Fig. 4.13). The combined effects on sawfly generation $t + 2$ of winter precipitation during sawfly generations t and $t + 1$ account for about 70 percent of the variance in gall density in dry-site clones. In wet-site clones, variation in precipitation in t accounts for 52 percent of the variation in gall density in $t + 1$, with no additional detectable effect of precipitation in sawfly generation $t + 2$ (Price *et al.* 1998a).

CONCLUSIONS

These results are remarkable for several reasons. First, it is most unusual to be able to account for 70 percent of the variation in population

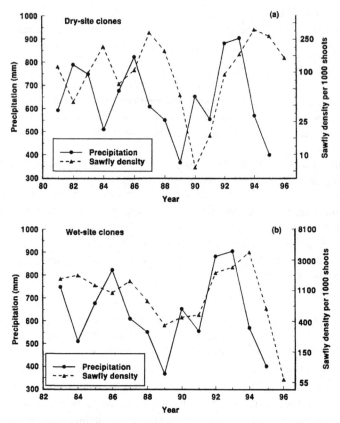

Fig. 4.11. Time-series analysis for 16 generations of *Euura lasiolepis* on dry-site clones (above) and for 14 generations on wet-site clones (below). Note the tendency for winter precipitation to cycle, followed by a delayed cycle of sawfly density. (From Hunter, M. D. and P. W. Price (1998) Cycles in insect populations, Fig. 2, *Ecol. Entomol.* 23: 216–222, Blackwell Science, Oxford.)

by a single factor, precipitation in this case. Second, it is remarkable that bottom-up forces through host plant variation can affect herbivore populations so dramatically in this system, overriding all other factors so commonly invoked as key factors in population dynamics. Third, cyclical population dynamics is explained by cyclical weather patterns, which set the carrying capacity for populations from the bottom up. Fourth, cyclicity has been explained traditionally by the delayed density dependence of natural enemies, but here we show that such cyclicity is accounted for by weather factors, which also have delayed effects. Fifth, we have explained the population dynamics of *Euura lasiolepis*

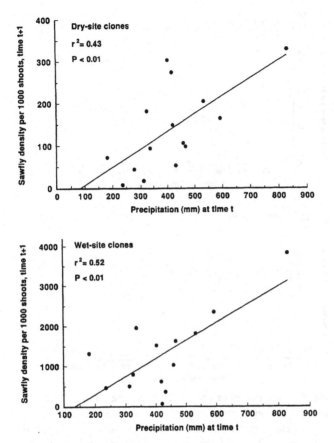

Fig. 4.12. Regressions of sawfly density in generation $t + 1$ in relation to precipitation in sawfly generation t, for dry-site clones (above) and for wet-site clones (below). (From Hunter, M. D. and P. W. Price (1998) Cycles in insect populations, Fig. 3, *Ecol. Entomol.* 23: 216–222, Blackwell Science, Oxford.)

mechanistically, using field observations over many years, intimately coupled with experiments on plant–animal interactions, plant module demographics, herbivore behavior, three-trophic-level interactions, chemical ecology, and physiological studies. Sixth, we have accounted for both proximate mechanisms (with tentative results only on causes of larval death) and ultimate mechanisms working on the distribution, abundance, and population dynamics of this stem-galling sawfly. Seventh, I suggest that there is not another species of any kind, in a more or less natural setting, that is as well understood in terms of population dynamics as *Euura lasiolepis*. A debatable point, no doubt, but one worth

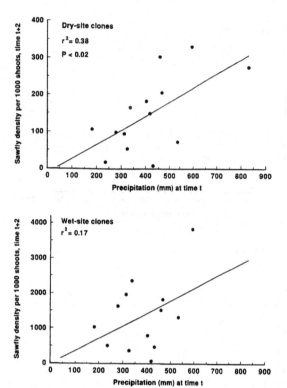

Fig. 4.13. Regressions of sawfly density in generation $t + 2$ in relation to precipitation in sawfly generation t, for dry-site clones (above) and for wet-site clones (below). (From Hunter, M. D. and P. W. Price (1998) Cycles in insect populations, Fig. 4, *Ecol. Entomol.* 23: 216–222, Blackwell Science, Oxford.)

examining closely. Admittedly, the system is very simple, once the essentials are understood, but with older methods of life table analysis, sampling methods, and preconceived notions of the important factors at play, we may well have reached as unconvincing a conclusion as in many other studies.

5

The focal group – The common sawflies

When one species is understood regarding distribution, abundance, and population dynamics, it is essential to test for the generality of the patterns discovered. "Study major, broad, repeatable patterns" in nature, admonished Tilman (1989, p. 90). "Because the purpose of ecology is to understand the causes of patterns in nature, we should start by studying the largest, most general, and most repeatable patterns." However, discovering broad repeatable patterns in nature offers a serious challenge, one that we have investigated for the last 15 years. In this chapter we discuss species related to *Euura lasiolepis* that have been studied by our research group. All are members of the common sawfly Family Tenthredinidae in the Order Hymenoptera. They form the focal group in which the patterns found for *Euura lasiolepis* are most likely to be repeated. And, if theory is the mechanistic explanation of broad patterns in nature, then finding such patterns in the family of common sawflies would provide the basis for theory. We have already described the mechanistic basis of pattern in distribution, abundance, and population dynamics in *Euura lasiolepis*. Do these patterns and mechanisms hold for a broad range of related species?

The family Tenthredinidae is composed mostly of free-feeding sawflies, but includes several gall-inducing genera and a few stem borers, leaf miners, fruit feeders, and catkin feeders (Table 5.1). The gall-inducing genera have received most of our attention and are restricted to the host plant family Salicaceae, with only two genera, *Populus*, the poplars and cottonwoods, and *Salix*, the willows. Species of *Euura* induce galls in stems, buds, petioles, and leaf midribs (Fig. 5.1). Another genus, *Pontania*, induces leaf galls, and the genus *Phyllocolpa* causes leaves to fold or roll at the edge (Figs. 5.2–5.4). We have investigated representatives of all genera and all types of gall formed by *Euura* species. In

Table 5.1. *Summary of the distribution of feeding types in the common sawflies, Family Tenthredinidae, given by the number of species recorded in each type in North America and Europe*

Total species	External leaf feeders	Stem borers	Leaf and bud miners	Fruit and catkin feeders	Leaf folders[a]	Typical gall formers
North America						
824	677	1	25	21	26	74
Europe						
795	698	8	34	18	12	25

[a] Leaf folders in the genus *Phyllocolpa* cause hypertrophy of the upper leaf lamina, causing the leaf edge to fold over, lower surface to lower surface, to form a gall, although uncharacteristic when compared to galls formed by *Pontania* and *Euura* species.
Source: Based on Price and Roininen (1993).

addition we have studied species across a broad longitudinal range in this predominantly Holarctic group.

OTHER *Euura* SPECIES

When studies on *Euura lasiolepis* were developing well we extended our concern to a bud galler, *Euura mucronata*, in the vicinity of Joensuu, North Karelia, in Finland. We quickly found that the patterns held for this species also (Price *et al.* 1987a, b). As willow ramets of the host plant species, *Salix cinerea*, aged, mean shoot length declined, the number of galls per shoot declined, and the percentage of buds galled declined (Fig. 5.5). As shoot length declined, bud gall diameter declined and survival of larvae declined. Presenting data on *Euura mucronata* for direct comparison with Fig. 3.6 on *Euura lasiolepis* shows a similar pattern of attack, with galls formed most frequently on rare long shoots (Fig. 5.6). Concentration of attack by the bud galler on rare, long shoots results in the same kind of distribution, abundance, and population dynamics as in the stem galler.

In fact, all species of *Euura* we have studied show the same pattern of attack on whichever willow is utilized as a host and wherever we have studied them – in the United States, Finland, and Japan (Fig. 5.7). A more general pattern emerges.

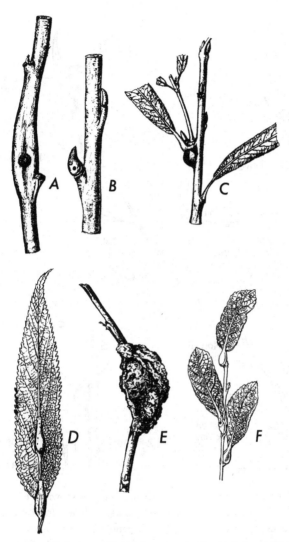

Fig. 5.1. Gall types induced by *Euura* species in Europe. (A) Stem gall by *Euura atra*. (B) Bud gall by *Euura mucronata*. (C) A transitional form between bud, petiole, and stem galls by *Euura laeta*. (D) Leaf midrib gall by *Euura testaceipes*. (E) Stem gall by *Euura amerinae*. (F) Petiole gall by *Euura venusta*. (From Pschorn-Walcher 1982.)

OTHER SAWFLIES

Adding other species and genera of sawflies, including gall inducers, free-feeders, and shoot borers, a large number of species show a positive, significant response to shoot length (Table 5.2) and 13 of these species

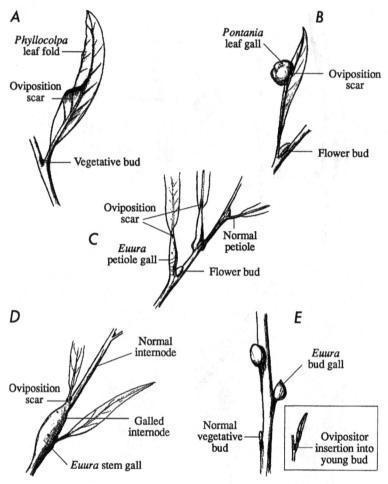

Fig. 5.2. Gall types induced by tenthredinid sawflies showing the position of oviposition scars relative to gall stimulation. (A) *Phyllocolpa* leaf fold. (B) *Pontania* leaf gall. (C) *Euura* petiole gall. (D) *Euura* stem gall. (E) *Euura* bud gall. Inset shows ovipositor insertion through the petiole of a very young leaf into an early bud to initiate a bud gall, as in *Euura mucronata*. (From Price and Roininen 1993.)

have been studied in enough detail to show strong plant quality effects in relation to an ovipositional preference linked to larval performance (Table 5.3).

Here we have a broad, general, repeatable pattern in nature and the mechanisms that drive that pattern. We have, then, the basis for a theory on the distribution, abundance, and population dynamics of tenthredinid sawflies on woody plants but heavily biased to host plants in the family Salicaceae.

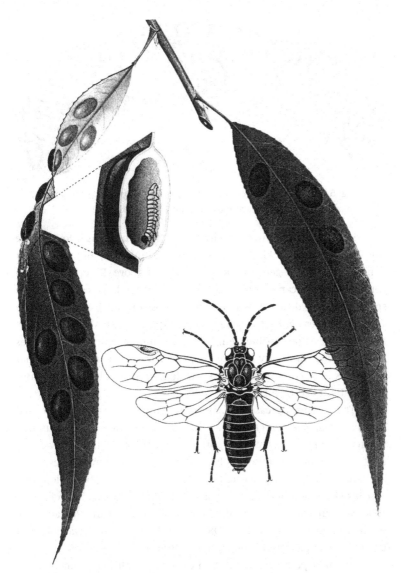

Fig. 5.3. Example of a gall-inducing sawfly, *Pontania proxima*, showing an adult female, galls on willow leaves, and a larva within a gall. (From Kopelke 1982.)

EXCEPTIONS

Where are the exceptions? What diminishes the validity of the general patterns and mechanisms we have established? Most of the cases we have studied show a pattern of attack on longer shoots. Long, vigorous shoots are generally uncommon or rare in a population of shoots,

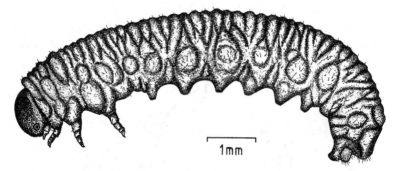

1mm

Fig. 5.4. Fifth instar larva of *Pontania proxima*, illustrating the general appearance of sawfly larvae in galls. (From Kopelke 1982.)

meaning that the carrying capacity for sawfly populations is low and we can expect the same kind of distribution, abundance, and population dynamics as in the general pattern discussed above. Exceptions to the general patterns can be grouped into three general classes: first, *Euura* species on trees differ in dynamics from those on shrubs; second, some *Pontania* species oviposit very early in the spring when shoots are short and difficult or impossible to discriminate among; third, some species and locations show very high or very low survival of larvae, such that no trends in survival relative to shoot length are evident. These cases will be discussed in turn.

The interesting case of *Euura* stem gallers on trees is illustrated by *Euura amerinae*, which attacks *Salix pentandra* in Europe. This tree species colonizes open mineral soil in Joensuu, Finland, where we conducted our study, forming highly patchy resources for sawflies. Galls are large, woody, and persistent (see Fig. 5.1E) so that populations can be censused accurately several years after attacks occur. Therefore, in 1986 we could detect the original colonization event in 1983 and subsequent population growth to 1986. Then the population was followed until extinction in 1991 (Roininen *et al.* 1993a).

The *Salix pentandra* population became established in about 1978. Tree height and mean shoot length had both increased and growth rates remained high during the demise of the sawfly population (Fig. 5.8).

Fig. 5.5. The willow–sawfly relationships between *Salix cinerea* and *Euura mucronata*, a bud-galling species near Joensuu, Finland. (A) As ramet age increases mean shoot length declines. (B) As ramets increase in age and shoot length declines, the mean number of galls per shoot declines. (C) As ramet age increases the percentage of all buds galled declines. (From Price *et al.* 1987a.)

A

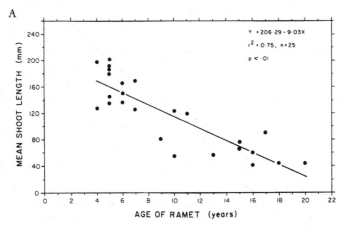

B

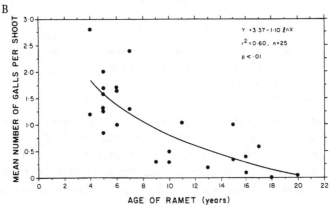

C

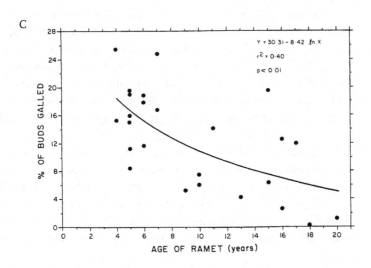

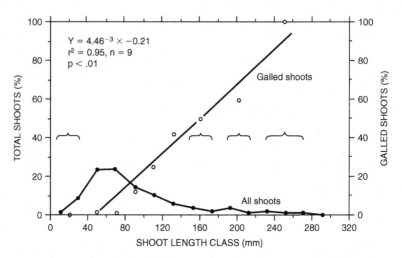

Fig. 5.6. The relationship between shoot length class on the willow, *Salix cinerea*, the percentage of shoots in each class, and the percentage of shoots galled by the bud-galling sawfly *Euura mucronata*. Note the similarity in patterns to those for *Salix lasiolepis* and *Euura lasiolepis* in Fig. 3.6. (From Price *et al.* 1987b.)

But local extinction of sawflies occurred when trees were only about 13 years old and in their prime. Evidence suggested that natural enemies were not important in this system, but productivity per gall declined monotonically with tree age. There appeared to be a maturational effect of host trees on the *Euura* population, or an ontogenetic aging effect.

The main difference between this *Euura* population on a tree and the others we have studied mostly on shrubs is that trees have apical dominance and shrubs do not. Trees produce a diminishing resource of long shoots after early vigorous growth because of physiological or ontogenetic aging (cf. Kearsley and Whitham 1989). In shrubs, the architecture is completely different, with new shoots developing from the base of the shrub, producing a more sustained resource for sawflies requiring vigorous growth on young ramets. Hence, we see rather stable dynamics on shrubs, as in Fig. 4.11, and a flush–crash cycle in the tree-dwelling population. In other respects *Euura amerinae* is very similar in its ecology to *Euura lasiolepis*. Females use a specific phenolic glucoside as an oviposition cue (Kolehmainen *et al.* 1994) and they show a strong preference for long shoots (Table 5.2). They remain at low abundance over a landscape and distribution is very patchy.

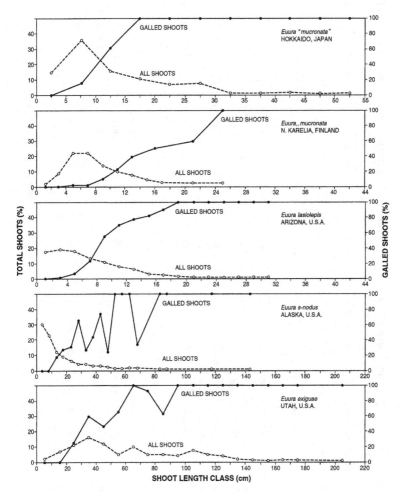

Fig. 5.7. Relationships between shoots available on willow hosts and the attack by various *Euura* species, in Japan, Finland, and the United States of America. Note the similarity of pattern of availability of shoots and percentage of shoots attacked. *Euura "mucronata"* in Hokkaido, Japan, is probably a different species from the European *Euura mucronata* because of the different host species utilized and the geographic distance involved. More erratic patterns of attack by *Euura s-nodus* and *Euura exiguae* result from very small populations of sawflies. (Based on Price *et al.* 1995a.)

Euura atra also attacks trees, such as *Salix alba* and *Salix fragilis*, but the only populations we could find were on ornamental trees that were frequently pruned, keeping shoots in a vigorous mode of growth. Otherwise, small numbers of galls were found on old trees heavily

Table 5.2. *Species of tenthredinid sawfly showing a positive, significant response to shoot length (with one exception). An additional one species of pamphiliid sawfly is listed as species number 36, and a cephid sawfly as species 38*

Insect species	Plant species	Locality	r^{2a}	p	Reference
Stem gallers					
1. *Euura amerinae*	*Salix pentandra*	Joensuu, Finland	0.92	<0.01	Price *et al.* 1990
2. *Euura atra*	*Salix alba*	Joensuu, Finland	0.83	<0.01	Price *et al.* 1997
3. *Euura "atra"* n. sp.	*Salix rosmartinifolia*	Turku, Finland	0.47	N.S.	H. Roininen unpublished data
4. *Euura exiguae*	*Salix exigua*	Weber River, Utah	0.70	<0.01	Price 1989
5. *Euura lasiolepis*	*Salix lasiolepis*	Flagstaff, Arizona	0.94	<0.001	Craig *et al.* 1986
6. *Euura* n. sp.	*Salix interior*	Urbana, Illinois	0.73	<0.01	Price *et al.* 1990
7. *Euura s-nodus*	*Salix interior*	Tanana River, Alaska	0.75	<0.01	P. W. Price and H. Roininen unpublished data
Bud gallers					
8. *Euura lappo*	*Salix lapponum*	Kevo, Finland	0.62	<0.01	Price *et al.* 1994
9. *Euura mucronata*	*Salix cinerea*	Joensuu, Finland	0.93	<0.01	Price *et al.* 1987a
10. *Euura "mucronata"*	*Salix sachalinensis*	Lake Shitose, Hokkaido, Japan	0.90	<0.01	P. W. Price and T. Ohgushi unpublished data
11. *Euura "mucronata"*	*Salix scouleriana*	San Francisco Peaks, Arizona	0.51	<0.003	Ferrier 1999
Petiole galler					
12. *Euura* n. sp.	*Salix lasiolepis*	Flagstaff, Arizona	0.63	<0.01	Price *et al.* 1990

	Host plant	Location	r	p	Reference
Midrib galler					
13. *Euura* n. sp.	*Salix exigua*	Lees Ferry, Arizona	0.69	<0.005	Woods *et al.* 1996
Leaf lamina gallers					
14. *Pontania amurensis*	*Salix miyabeana*	Sapporo, Hokkaido, Japan	0.67	<0.05	Price *et al.* 1999
15. *Pontania* nr. *pacifica*	*Salix lasiolepis*	Flagstaff, Arizona	0.53	<0.01	Fritz and Price 1988
16. *Pontania pustulator*	*Salix phylicifolia*	Rantakylä, Finland	0.80	<0.01	Price *et al.* 1999
17. *Pontania* "sp. 1"	*Salix scouleriana*	Flagstaff, Arizona	0.58	<0.05	Price *et al.* 1999
Leaf edge folders and rollers					
18. *Phyllocolpa coriacea*	*Salix cinerea*	Joensuu, Finland	0.86	<0.01	Price *et al.* 1990
19. *Phyllocolpa excavata*	*Salix pentandra*	Joensuu, Finland	0.51	<0.01	Price *et al.* 1990
20. *Phyllocolpa* sp. folder	*Salix lasiolepis*	Flagstaff, Arizona	0.87	<0.01	Price *et al.* 1990
21. *Phyllocolpa* sp. roller	*Salix lasiolepis*	Flagstaff, Arizona	0.73	<0.01	Price *et al.* 1990
22. *Phyllocolpa* sp.	*Salix miyabeana*	Sapporo, Hokkaido, Japan	0.76	<0.002	Price and Ohgushi 1995
23. *Phyllocolpa bozemani*	*Populus tremuloides*	Flagstaff, Arizona	0.70	<0.01	Price *et al.* 1990
24. *Phyllocolpa* sp.	*Populus balsamifera*	Tanana River, Alaska	0.55	<0.02	Roininen *et al.* 1997
25. *Phyllocolpa* sp.	*Salix novae-angliae*	Tanana River, Alaska	0.59	<0.01	Roininen *et al.* 1997
26. *Phyllocolpa leavittii*	*Salix discolor*	Milford, New York	0.89	<0.001	Fritz *et al.* 2000
Free feeders					
27. *Nematus iridescens*	*Populus tremuloides*	Flagstaff, Arizona	0.95	<0.01	Price and Carr 2000
28. *Nematus oligospilus*	*Salix lasiolepis*	Flagstaff, Arizona	0.78	<0.01	Price and Carr 2000
29. *Nematus vancouverensis*	*Populus tremuloides*	Flagstaff, Arizona	0.77	<0.01	Price and Carr 2000
30. *Nematus ferrugineus* group	*Salix myrsinifolia*	Parikkala, Finland		<0.001[b]	H. Roininen and P. W. Price unpublished data
31. *Eitelius dentatus*	*Salix phylicifolia*	Enonkoski, Finland		<0.001[b]	H. Roininen pers. com.

Table 5.2. (cont.)

Insect species	Plant species	Locality	r^{2a}	p	Reference
32. *Pristiphora forsiusi*	*Salix pentandra*	Parikkala, Finland		<0.01[b]	H. Roininen pers. com.
33. *Amauronematus eiteli*	*Salix pentandra*	Parikkala, Finland		<0.01[b]	H. Roininen pers. com.
34. *Amauronematus distinguensis*	*Salix pentandra*	Parikkala, Finland		<0.01[b]	H. Roininen pers. com.
35. *Decanematus* sp.[c]	*Salix pentandra*	Parikkala, Finland			H. Roininen pers. com.
36. *Pamphilus gyllenthali* Pamphiliidae	*Salix caprea*	Petkeljärvi, Finland		<0.01[b]	H. Roininen pers. com.
Shoot borers					
37. *Ardis brunniventris*	*Rosa pimpinellifolia*	Joensuu, Finland	0.99	<0.01	H. Roininen and P. W. Price unpublished data
38. *Cephus cinctus* Cephidae	*Triticum aestivum*	Pondera County, Montana			Morrill *et al.* 2000

[a] r^2 values are for the regression of probability of attack per shoot length class on shoot length.

[b] *t*-test results comparing length of attacked and unattacked shoots per plant. Most host plants were young, vigorous plants 2–5 years old. I am very grateful to Dr. Heikki Roininen for these data.

[c] This species attacks main shoots, stunting their growth, so these otherwise vigorous shoots soon become shorter than others. Hence, shoot length data are misleading.

Table 5.3. *Thirteen studies on tenthredinid sawflies showing strong plant quality effects on ovipositional preference linked to larval performance*

Species	Feeding type	Location	Reference
1. *Euura* sp. 1	Midrib gall	Lees Ferry, Arizona	Woods *et al.* 1996
2. *Euura* sp. 2	Petiole gall	Flagstaff, Arizona	Stein and Price 1995
3. *Euura amerinae*	Stem gall	Joensuu, Finland	Roininen *et al.* 1993a
4. *Euura atra*	Stem gall	Joensuu, Finland	Price *et al.* 1997
5. *Euura exiguae*	Stem gall	Weber River, Utah	Price 1989
6. *Euura lasiolepis*	Stem gall	Flagstaff, Arizona	Craig *et al.* 1989
7. *Euura mucronata*	Bud gall	Joensuu, Finland	Price *et al.* 1987b, c
8. *Phyllocolpa* sp.	Leaf edge gall	Sapporo, Japan	Price and Ohgushi 1995
9. *Phyllocolpa leavitii*	Leaf edge gall	Milford, New York	Fritz *et al.* 2000
10. *Pontania* nr. *pacifica*	Leaf lamina gall	Flagstaff, Arizona	Stein and Price 1995
11. *Pontania* sp.	Leaf lamina gall	Flagstaff, Arizona	Price *et al.* 1999
12. *Nematus oligospilus*	Free feeding	Flagstaff, Arizona	Carr *et al.* 1998
13. *Nematus iridescens*	Free feeding	Flagstaff, Arizona	Carr 1995

damaged by snow, such that new vigorous shoots sprouted near the wounds (Price *et al.* 1997). Under natural conditions, undisturbed by humans, the emergent properties of *Euura atra* on tree hosts would undoubtedly be equivalent to those of *Euura lasiolepis*.

The second kind of exception to the general pattern is in species that emerge early in the spring. We have not studied these species in detail yet but the pattern is clear enough. It is observed in the *Pontania viminalis* group in Finland (H. Roininen, pers. comm.) and in an undescribed species of *Pontania* in Japan, near Sapporo on the island of Hokkaido. The Japanese species attacks *Salix sachalinensis* along the Ishikari River and galls are clustered on the basal leaves of shoots, whether the shoots are long or short (Fig. 5.9). This pattern of attack is distinctly different from another *Pontania* species and a cecidomyiid gall midge species on the same trees and shoots. *Pontania aestiva* on *Salix phylicifolia* near Joensuu, Finland showed a pattern of attack very similar to that in Fig. 5.9.

The third group of species not conforming to the general pattern involves those that show a strong positive response to shoot length

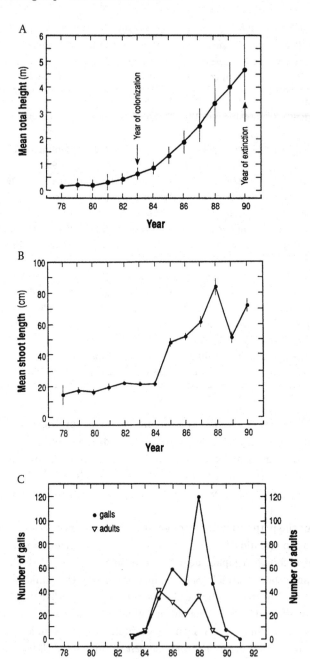

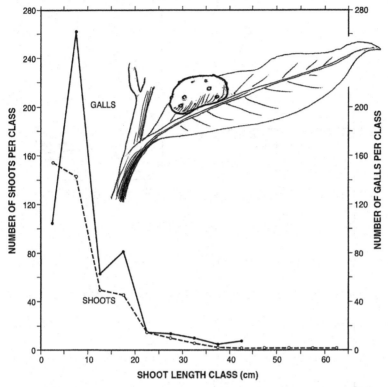

Fig. 5.9. The distribution of *Pontania* leaf galls on *Salix sachalinensis* at one site on the Ishikari River, near Sapporo, Hokkaido, Japan. Note the high number of galls on the shortest shoot length classes. The species of gall inducer is apparently not described (Yukawa and Masuda 1996) so an illustration of the yellow gall is included.

but little or no correlation with larval survival. Establishment of larvae in galls, a critical phase in the life cycle, may be uniformly high or low across all shoot length classes and ultimate survival in the gall may show similar patterns (Fig. 5.10). These kinds of relationships have

Fig. 5.8. (Opposite) The interaction between *Salix pentandra* host plants and the stem-galling sawfly *Euura amerinae*. (A) Trees established in a disturbed site in about 1978 and grew in height to about 5 m by 1990. (B) Mean shoot length increased and remained high from 1985 to 1990. (C) The population of stem galls and emerging adult sawflies rose and declined rapidly but never numbered more than 40 adults. (From Roininen, H., P. W. Price, and J. Tahvanainen (1993) Colonization and extinction in a population of the shoot-galling sawfly, *Euura amerinae*, Fig. 1, *Oikos* 68: 448–454, Munksgaard, Copenhagen.)

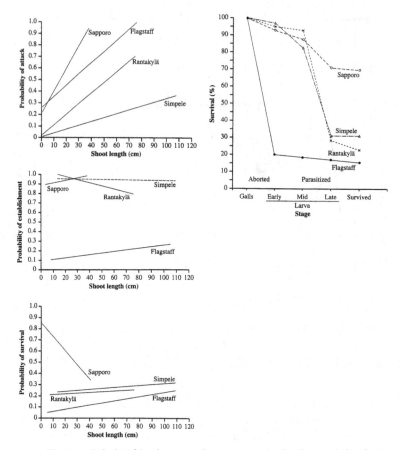

Fig. 5.10. Relationships between three *Pontania* leaf gallers and the shoot
lengths of their respective host plants in Japan, Finland, and the United
States, and survivorship curves for the different species and sites.
Pontania amurensis on *Salix miyabeana* occurred near Sapporo, Japan.
Pontania pustulator was on *Salix phylicifolia* in eastern Finland, near
Joensuu, studied at two sites, Rantakylä and Simpele. Near Flagstaff,
Arizona, an undescribed *Pontania* species was found on *Salix scouleriana*.
On the left are shown, in relation to shoot length class, the probabilities
of attack, the probabilities of larvae establishing a feeding site in the
gall, and the probability of survival to emergence of larvae from the
gall. On the right, the survivorship curves show the timing and extent
of death in the gall. (From Price *et al.* 1999.)

been found in three species of *Pontania* (Fig. 5.10), one species of *Euura*
(Ferrier 1999), and one free-feeding sawfly, *Nematus vancouverensis* (Carr
1995). Evidently, there are advantages for females to oviposit into vigorous modules independent of larval survival. Various hypotheses may

Table 5.4. *Known exceptions in tenthredinid sawflies to the general positive response by ovipositing females to long shoots*

Insect species	Plant species	Locality	Pattern of attack
1. *Pontania arctica*	*Salix phylicifolia*	Joensuu, Finland	Oviposits on early leaves independent of shoot length.
2. *Pontania aestiva*	*Salix myrsinifolia*	Joensuu, Finland	Oviposits on early leaves. Probably random attack when all shoots have a few leaves available.
3. *Pontania* sp.	*Salix sachalinensis*	Sapporo, Japan	See Fig. 5.9. On early leaves of all shoots.
4. *Nematus pavidus*	*Salix caprea, myrsinifolia, phylicifolia*	Joensuu, Finland	Random on all host species.

Source: H. Roininen and P. W. Price, unpublished data.

be relevant and have yet to be resolved. Target size is larger on more vigorous shoots, especially relevant to a *Euura* bud galler that oviposits into tiny buds early in bud development. Tissues on vigorous shoots are likely to be more tender, causing less wear on the ovipositor and costing less energy for oviposition. Or, in other parts of the landscape, attack on long shoots may have a direct benefit on larval survival. We have also recorded higher levels of phenolic glucosides in longer shoots which act as oviposition stimulants, suggesting the simple relationship between strength of ovipositional stimulant and probability of attack (cf. Price *et al.* 1989; Roininen *et al.* 1999).

Even though these deviations from pattern exist, most species, except those in the second kind of exception, have evolved or inherited a preference for oviposition into long, vigorously growing shoots (see Table 5.4 for all known exceptions). This is the basic component of the adaptive syndrome that sets limits on the emergent properties, as stated early in Chapter 4. The limits are shortage of resources in time and space, because long shoots are scarce. In the case of tree-dwelling gall-inducing sawflies these limits are compounded by the shortage of time a tree and a population of young trees are available as an adequate resource. Another generality common to the species studied is that the fundamental essentials of the population ecology are based on strong bottom-up effects through resources, and that natural enemies play at least a minor role in distribution, abundance, and population dynamics of these species.

ADAPTIVE RADIATION AND THE PHYLOGENETIC
DEVELOPMENT OF EMERGENT PROPERTIES

With the pattern of female sawflies ovipositing in longer shoots being repeated many times and showing a strong preponderance in the
species studied – 90 percent of species listed in Tables 5.2 and 5.4 – we
have evidence for strong repeatable patterns in nature. The general pattern is macroevolutionary because it covers many species, many different microsites utilized for gall induction, and many genera of sawflies:
all three genera of gall-inducing sawflies and seven genera of free-
feeding sawflies. This macroevolutionary pattern is explained mechanistically by the flow of evolutionary consequences stemming from the
plesiomorphic phylogenetic constraint of a plant-piercing ovipositor.
And the general ecological pattern emerging from this constraint and
adaptive syndrome is that species are patchily distributed on a landscape, confined to vigorous plant modules, they are usually low in
abundance, and their population dynamics tend to be stable relative to
eruptive or outbreak species. We have the basis for a macroevolutionary
explanation for macroecological patterns.

We should then inquire about the existence of additional patterns
at a broader scale, across the adaptive radiation of the sawflies. For
this we need to understand how the radiation developed. The general
scenario was hypothesized independently by Roininen (1991) and Price
(1992a; see also Price and Roininen 1993) and has been tested by Nyman
et al. (1998, 2000). Tommi Nyman and Heikki Roininen were generous
in calling this the Price–Roininen Hypothesis on the adaptive radiation
of gall-inducing sawflies, for publication precedence was Roininen's,
even though I had written the chapter with the hypothesis in 1982,
which was delayed in publication until 1992. The hypothesis was depicted by Nyman et al. (1998) (Fig. 5.11) and is embellished slightly here.
Free-feeding nematine sawflies radiated on members of the Salicaceae,
willow (Salix) and poplar (Populus), with females laying eggs into pockets cut under the leaf epidermis with a saw-like ovipositor. Today, we
see species that induce small swellings at the site of oviposition and
first instar larvae may even feed within such procecidia, or insipient
galls, as in the case of Amauronematus eiteli (H. Roininen, pers. comm.).
Females evolved with the capacity to modify host plant cell division
and development with fluid injected from a gland associated with the
ovipositor. Free feeders began to prick the leaf repeatedly near the leaf
margin, inducing swelling of tissues and a leaf fold or roll in which
the larva fed in a protected site. This repeated oviposition in the leaf
blade near the edge moved toward the midrib, causing sausage-shaped

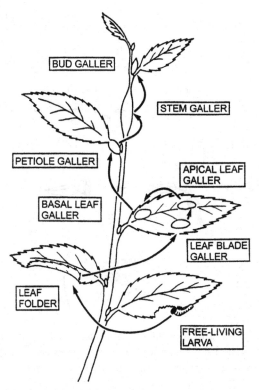

Fig. 5.11. The hypothesis that gall-inducing nematine sawflies evolved from free feeders, moving from leaf-edge gallers to galling the leaf blade, petiole, stem, and bud. (From Nyman *et al.* 1998.)

galls along the leaf lamina, with larvae feeding within the gall. Following this shift, round or oblong leaf blade galls developed, some more apically placed, some more basal, some expanding through both leaf surfaces, some only from the adaxial surface. Gall types and their hypothetical radiation are shown in Fig. 5.12 (cf. Figs. 5.1 and 5.2). From the basal leaf-gall type oviposition moved down the shoot axis into the petiole, stem, and bud. In addition, and perhaps first, oviposition into the leaf midrib produced midrib galls instead of leaf lamina galls closely associated with the midrib. When one appreciates that oviposition by the females inducing galls with truly endophytic larvae is accomplished when leaves are very young and small, it becomes clear that slight accidental shifts in oviposition, small inaccuracies by perhaps no more than 1 mm from the usual target, could induce a gall on a new substrate. This would constitute a shift into a new adaptive zone, followed by speciation when shifts to new host plant species occurred.

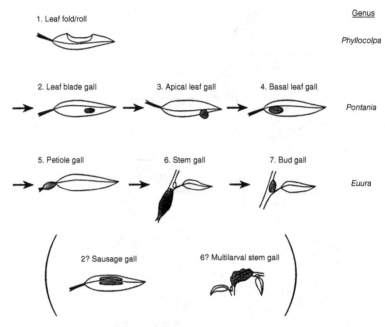

Fig. 5.12. The various gall types induced by nematine sawflies on willows with the hypothetical shifts from leaf edge toward the stem axis. The genera of sawflies are given on the right. (From Nyman 2000a.)

The adaptive radiation has clearly proceeded as a sequence of breakthroughs into new adaptive zones – leaf folds, leaf galls, stem galls, bud galls – with subsequent speciation during host plant shifts in each adaptive zone, as Roininen (1991) hypothesized (Fig. 5.13). Host shifting was more or less opportunistic based on ecological access to willow species associated with a current host (Fig. 5.14). No evidence exists of coevolution, cospeciation, phylogenetic tracking of hosts by sawflies, or host shifting among hosts with similar phytochemical profiles (Nyman 2000a).

The most recent hypothetical phylogeny of gall-type evolution is consistent with much of the Price–Roininen hypothesis (Nyman *et al.* 2000). Incorporating 31 gall-inducing sawfly species and an outgroup of five free-feeding nematine sawflies (Table 5.5), the tree progresses from free feeders to leaf folders and rollers, to leaf-blade gallers, and down the shoot axis to stem, bud, and petiole gallers (Fig. 5.15). The macroevolutionary pattern of adaptive radiation in the gall-inducing sawflies is evident.

Are there patterns in emergent properties that follow the macro-evolutionary pattern, and are they the obvious consequence of the

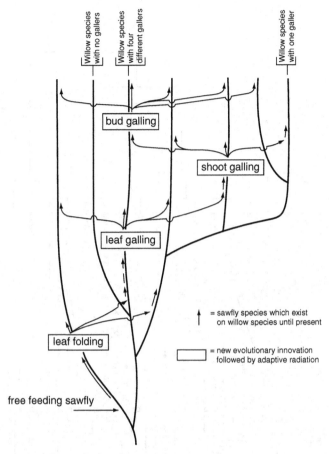

Fig. 5.13. Roininen's (1991) hypothesis on the radiation of gall-inducing sawflies on willows, with each new gall type having a unique origin and then with species spreading to new willow hosts. The figure illustrates the diversification of willows followed by sawflies without any coevolution, cospeciation, or tracking in the lineage.

phylogenetic trend? There is evidence of increasing specificity along the phylogeny from free feeders to stem gallers. Some *Nematus* species, such as *Nematus pavidus*, feed on at least three host species even though all are willows, while stem gallers are known mostly to attack only one host species. Where multiple host species have been recorded, as for *Euura atra* and *Euura mucronata*, the trend has been to discover cryptic, sibling species complexes, with high host specificity for each species (e.g. Roininen *et al.* 1993b; Kopelke 2000, 2001; Nyman 2000b). In addition the range of module quality appears to become more limiting from free feeders to stem gallers. Some *Nematus* attack shoot lengths at random

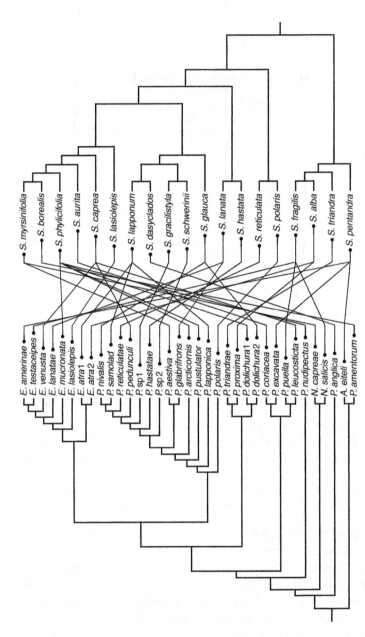

Fig. 5.14. The host association of nematine sawflies and the phylogenies of willow hosts on the left and sawflies on the right. Full names of species are given in Table 5.5. Note the lack of any concordance among willows and sawflies. (From Nyman 2000a.)

Table 5.5. *Gall types, sawfly species names, host plant willow species, and sites of collection for the 36 sawfly species used in the construction of the phylogenetic hypothesis by Nyman et al. (2000)*

Gall type[a]	Species	Willow host[b]	Sample site,[c] date
0. Outgroup	*Tenthredo arcuata* (Förster)	—	Joensuu (F), 1996
	Nematus salicis (Linnaeus)	*S. fragilis* (S)	Joensuu (F), 1996
	Nematus capreae (Linnaeus)	*S. myrsinifolia* (V)	Kilpisjärvi (F), 1997
	Pontopristia amentorum (Förster)	*S. borealis* (V)	Kilpisjärvi (F), 1997
	Amauronematus eiteli (Saarinen)	*S. pentandra* (S)	Parikkala (F), 1998
1. Leaf fold/ roll (11)	*Phyllocolpa nudipectus* (Vikberg)	*S. phylicifolia* (V)	Joensuu (F), 1996
	Phyllocolpa leucosticta (Hartig)	*S. caprea* (V)	Puhos (F), 1997
	Phyllocolpa puella (Thomson)	*S. fragilis* (S)	Joensuu (F), 1997
	Phyllocolpa excavata (Marlatt)	*S. pentandra* (S)	Joensuu (F), 1997
	Phyllocolpa coriacea (Benson)	*S. aurita* (V)	Mekrijärvi (F), 1997
	Phyllocolpa anglica (Cameron)	*S. dasyclados* (V)	Krasnojarsk (R), 1993
2. Leaf blade gall (3–5)	*Pontania proxima* (Lepeletier)	*S. alba* (S)	Joensuu (F), 1991
	Pontania triandrae (Benson)	*S. triandra* (S)	Keminmaa (F), 1997
2? Sausage gall (4–10)	*Pontania dolichura* (Thomson)	*S. phylicifolia* (V)	Paanajärvi (R), 1996
	Pontania dolichura (Thomson)	*S. glauca* (C)	Kilpisjärvi (F), 1997
3. Apical leaf gall (18)	*Pontania glabrifrons* (Benson)	*S. lanata* (V)	Kanin Peninsula (R), 1994
	Pontania samolad (Malaise)	*S. lapponum* (V)	Paanajärvi (R), 1996
	Pontania reticulatae (Malaise)	*S. reticulata* (C)	Kolguyev Island (R), 1994
	Pontania pedunculi (Hartig)	*S. caprea* (V)	Lebed-Ozero (R), 1996

Table 5.5. (cont.)

Gall type[a]	Species	Willow host[b]	Sample site,[c] date
	Pontania nivalis (Vikberg)	S. glauca (C)	Kakhovskiy Bay (R), 1994
	Pontania hastatae (Vikberg)	S. hastata (V)	Björkstugan (S), 1989
	Pontania arcticornis (Konow)	S. phylicifolia (V)	Paanajärvi (R), 1996
	Pontania sp.	S. schwerinii (V)	Ussuri Reserve (R), 1996
	Pontania sp.	S. gracilistyla (V)	Kedrovaja Pad Reserve (R), 1996
	Pontania aestiva (Thomson)	S. myrsinifolia (V)	Paanajärvi (R), 1996
4. Basal leaf gall (8)	Pontania lapponica (Malaise)	S. lapponum (V)	Kilpisjärvi (F), 1997
	Pontania pustulator (Forsius)	S. phylicifolia (V)	Joensuu (F), 1994
	Pontania polaris (Malaise)	S. polaris (C)	Kilpisjärvi (F), 1997
5. Petiole gall (2–3)	Euura testaceipes (Zaddach)	S. fragilis (S)	Joensuu (F), 1989
	Euura venusta (Zaddach)	S. caprea (V)	Härskiä (F), 1989
6. Stem gall (4–8)	Euura atra (Jurine)	S. alba (S)	Simpele (F), 1988
	Euura atra (Jurine)	S. lapponum (V)	Kilpisjärvi (F), 1997
	Euura lasiolepis (Smith)	S. lasiolepis (V)	Flagstaff, Arizona (U), 1997
6? Multilarval stem gall (1)	Euura amerinae (Linnaeus)	S. pentandra (S)	Joensuu (F), 1998
7. Bud gall (3–11)	Euura mucronata (Hartig)	S. phylicifolia (V)	Kilpisjärvi (F), 1997
	Euura lanatae (Malaise)	S. lanata (V)	Kilpisjärvi (F), 1997

[a] Numbers in parentheses indicate numbers of European species according to Price and Roininen (1993), Kopelke (1994), and personal observation. The numbers are only approximate because the status of many currently recognized species is uncertain.

[b] Letters in parentheses indicate the willow subgenus to which the host belongs: S, Salix; C, Chamaetia; V, Vetrix.

[c] F, Finland; R, Russia; S, Sweden; U, United States.

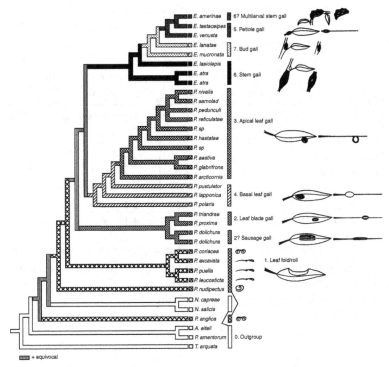

Fig. 5.15. A phylogenetic hypothesis of gall type evolution in nematine sawflies and a test of the Price–Roininen Hypothesis. Full names of species are provided in Table 5.5. Illustrations of gall types and cross-sections through galls are provided on the right. (From Nyman et al. 2000.)

as in *Nematus pavidus*. Some *Pontania* can attack and survive well on a wide range of leaf sizes, as in *Pontania* sp. near *pacifica* (Clancy *et al.* 1993) and at high population densities the pattern of attacking large leaves on long shoots is lost.

Many *Pontania* galls can be produced on a single leaf, up to ten or more in some cases (Clancy *et al.* 1986; cf. Fig. 5.3), resulting in the potential for high population densities, much higher than in *Euura* stem gallers in which only one gall can develop per node. Another trend is the potential for high mortality in free feeders to leaf gallers, caused by natural enemies, while in *Euura* species such mortality is generally much lower (Price and Pschorn-Walcher 1988) (Table 5.6). However, plant resistance factors are reduced in free feeders, leaf folders, and leaf gallers and increase as galls are formed on stems and buds. All these trends need more study for, at this time, they are based on familiarity with only a few species on which detailed studies have been undertaken.

Table 5.6. *Summary showing the decline in the number of species of parasitoid attacking nematine sawflies as the sawflies become more concealed in galls and the corresponding decline in mortality inflicted*

Sawfly feeding type	Mean number of parasitoid species			Mean range in mortality (%)[b,c]		
	Larval	Eonymphal[a]	Cocoon	Larval	Eonymphal	Cocoon
Colonial exposed feeders (n = 13)	5.5	1.3	9.0	3–49	3–29	1–24
Solitary exposed feeders (n = 11)	4.2	3.0	6.0			
Leaf gallers (n = 11)	5.5	1.0	1.0	22–57	3	
Shoot and bud gallers (n = 5)	4.0	0²	0	4–13	0	0

[a] The eonymph is a nonfeeding last instar larva which has specialized parasitoids in the free-feeding sawfly, not found on stem gallers.
[b] Zero values indicate no parasitoid species present.
[c] Blank space indicates lack of information.
Source: Price and Pschorn-Walcher (1988).

These patterns can be summarized for gall types in relation to the phylogenetic hypothesis, concentrating on the central emergent properties of distribution, abundance, and population dynamics (Fig. 5.16). Salient features in the adaptive radiation of gall-inducing sawflies and their ultimate effects on emergent properties are provided. Starting with free-feeding sawflies (1 in Fig. 5.16) some so-called free feeders actually initiate a procecidium in which a first instar larva feeds (2) indicating the potential for gall induction. Full gall initiation starts with a female initiating swelling of tissue along a leaf edge to form a leaf fold or roll in which the larva feeds (3). Noting that "Galls are defined as any deviation in the normal pattern of plant growth produced by a specific reaction to the presence and activity of a foreign organism (animal or plant)" (Dreger-Jauffret and Shorthouse 1992, p. 8), these leaf folds qualify fully as galls. Moving from leaf-edge galling to leaf-blade galls (4) may have taken two paths. A small shift in oviposition toward the leaf midrib, a retention of multiple injections of gall-inducing fluid, and a small increase in the stimulus would yield a sausage-shaped gall of the *Pontania dolichura* type (5), followed by reduction of gall length to leaf-blade galls of the *Pontania proxima* type. Alternatively, there may have been a direct move from leaf folding to

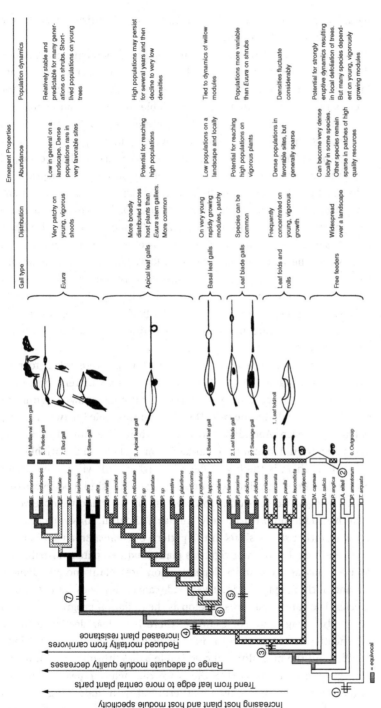

Fig. 5.16. A summary of emergent properties in the gall-inducing sawflies including general trends on the left and synopses of distribution, abundance, and population dynamics on the right. The numbered points on the phylogeny are described in the text. The phylogeny is from Nyman et al. (2000).

basal leaf galls (6), and from here to *Euura* species utilizing midribs and petioles of leaves, buds, and stems (7).

Concerning distribution, abundance, and population dynamics, descriptions of patterns are based on a small number of species we know adequately, and refer to gall types rather than directly to all the species listed in Fig. 5.16. We have available two studies on population dynamics, on *Euura lasiolepis* (e.g. Hunter and Price 1998), and on *Euura amerinae* (Roininen *et al.* 1993a), so obviously more quantitative studies are necessary. However, over the past 20 years we have collected intermittently density estimates of other sawflies around Flagstaff on *Salix lasiolepis* showing great differences in *Pontania* and *Euura* dynamics. For example, from 1980 to 1985 *Pontania* sp. near *pacifica*, an apical leaf gall type, declined from 4512 galls/1000 shoots to 107 (cf. Clancy *et al.* 1986) and has remained at low levels ever since. But on the same clones over the same period the stem galler *Euura lasiolepis* populations varied from 700 to 44 galls/1000 shoots (Price 1992b), the petiole galler *Euura* sp. remained between 398 and 96 galls/1000 shoots, and *Phyllocolpa* sp. stayed between 997 and 298 galls/1000 shoots. These data show that *Pontania* can reach much higher densities than *Euura* or *Phyllocolpa* species on the same willow clones.

When three experts on sawflies in Europe were asked to rank the species in Table 5.5 and Fig. 5.16 from 1 to 5 in relation to distribution, abundance, and dynamics, their rankings showed significant correlations in all but one case. Distribution was scaled between 1, very patchy to 5, even. Abundance was ranked from 1, common to 5, rare, and population dynamics was ranked from 1, stable to 5, variable.

Here, then, is the beginning of a phylogenetic view of emergent properties of ecological importance, the first of its kind, I believe. Based on a phylogenetic constraint in common for the clade we observe an adaptive radiation with broadly similar ecological consequences and yet with rather subtle shifts in the pattern of trends. Broad similarities are seen in exploitation of vigorous plant modules, patchy distribution over a landscape, dynamics dependent on host plant module dynamics, and nothing approaching the broad-scale outbreak dynamics of pest species. The subtler patterns develop as the radiation progresses, shown as trends in Fig. 5.16. For example, there is a clear decline in natural enemies in terms of both number of parasitoid species attacking sawflies and the mortality inflicted (Table 5.6). Within the ambit of the Phylogenetic Constraints Hypothesis we can detect repeated patterns in nature, adding up to a broad pattern, but with underlying patterns within the adaptive radiation based on clear progressions from one adaptive zone to another.

No doubt, there are other broad patterns which remain to be detected. One set of properties we did predict, on host plant use by sawflies, is treated in the next section of this chapter.

BIOGEOGRAPHIC PATTERNS

Because of the particular limits set by tree architecture and development, we predicted that we should find no more, or fewer, galling sawfly species on trees than on shrubs, quite contrary to the general patterns of increased insect herbivore species richness from shrubs to trees (e.g. Lawton and Schroder 1977; Lawton 1983). Shrubs lack apical dominance, usually providing a succession of new juvenile shoots year after year and remaining available to colonization by sawflies over many years or decades. Trees with apical dominance and the inevitable physiological and ontogenetic aging usually provide a declining supply of long, vigorous shoots over the years, resulting in reduced probability of colonization by sawflies (Price et al. 1998b).

The general pattern for insect herbivore species richness is to increase as the geographic range of the host plant species increases, and to increase as the architectural stature of the plant type increases from herbs to shrubs to trees (Lawton and Schroder 1977, 1978; Strong and Levin 1979; Strong et al. 1984) (Fig. 5.17). Although the mechanisms have not been tested adequately, the pattern is driven from the bottom up by the allocation of plant resources to growth. In the rare case where effects of competitors and natural enemies have been evaluated, the only valid effect is resource driven with no measurable lateral or top-down influences (Lawton and Price 1979).

The pattern for gall-inducing sawflies in relation to geographic range and architecture is significantly different from the general trends, and in the direction that we predicted. My friends Heikki Roininen and Alexei Zinovjev compiled a wonderful record of all known galling sawflies on all willow species in the Palaearctic Biogeographic Realm, and used Skvortsov's (1968) data on host geographic range (now available in English translation: Skvortsov 1999). The advantage of using only one plant genus and one taxon of herbivores is that chemical defenses, resources provided by the plant, and the adaptive syndrome of herbivores are more similar than when many taxonomic groups are involved in the analysis. Among the willow species only geographic range and architecture change. Lawton (1983) proposed two major hypotheses to account for the general pattern found outside the sawflies: the size per se hypothesis and the resource diversity hypothesis. All willows provide the same kinds of resources, all being woody plants. Be they dwarf

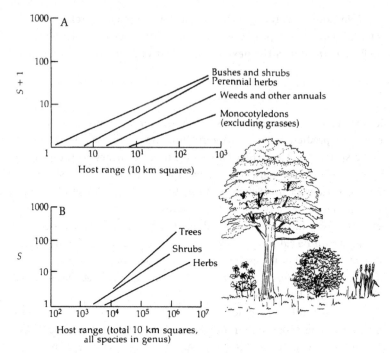

Fig. 5.17. The general relationships between plant host geographic range and the architecture of the hosts, showing that the larger plant types are colonized by more insect herbivore species at any given geographic range. (A) is based on Lawton and Schroder (1977), (B) is from Strong and Levin (1979), and the whole Figure is from Strong et al. (1984). (Drawing by John Lawton.) (Reproduced from Strong, D. R., J. H. Lawton, and T. R. E. Southwood (1984) *Insects on plants: Community patterns and mechanisms*, Fig. 3.7, Blackwell Science, Oxford.)

shrubs, large shrubs, or trees, all provide bark, stems, leaves, catkins, and seeds, so the resource diversity hypothesis can be excluded as a major influence on species richness per host plant. Therefore, we have a purer test of the size per se hypothesis than in past publications.

We showed that willow plant height, which ranged from a meter or less to 37 m, provided no explanatory power on the number of gall-inducing sawfly species per host plant species (Price et al. 1998c) (Fig. 5.18). The trends in species richness were more or less flat, but

Fig. 5.18. The number of species of gall-inducing sawflies recorded from willows of different architecture and height, for willows with large, moderate, and small geographic ranges in the Eurasian Biogeographic Realm. (From Price et al. 1998c.)

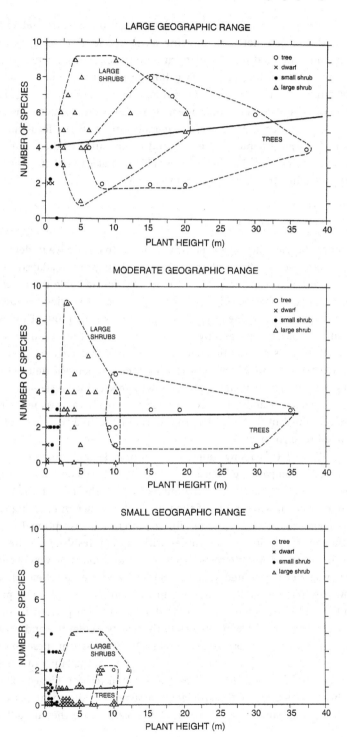

in all classes of geographic range, from large to moderate and small, some shrubs had more gall-inducing sawfly species than all trees with a much greater stature. This pattern is consistent with our hypothesis that trees are hard to colonize by gall inducers in evolutionary time because they lack the persistently rejuvenating new ramets coming from the base of a plant that keeps adequate resources available for sawflies over the long term. As a result of detailed study on the focal species *Euura lasiolepis* (Chapters 3 and 4), and comparative studies on many related species (this chapter), we were able to make a prediction on the biogeographic distribution of sawfly species.

NOTABLE POINTS

Perhaps the most important point to make is that once we understood one species reasonably well and patterns relating to the adaptive syndrome and emergent properties, we devoted considerable resources to comparative studies on other species, expanding the species considered as extensively as possible given time and funding constraints. This comparative approach was greatly simplified by noting the significance of shoot length as a synoptic character with closely correlated traits of growth rate, ultimate length, leaf size, internode length, stem width, bud size, and phytochemistry. Some sawfly species may respond to one set of characters and others to another set, but the aggregate characters contained in the shoot length criterion permitted the detection of a broad pattern in nature.

We feel that many more studies would benefit from this kind of broadly comparative approach to population ecology, for the discovery of broad patterns is fundamental to the theory of nature. We emphasized the importance of the weight of evidence in reaching general conclusions while maintaining an interest in exceptional cases. Although publishing more studies with similar results on different species meets with some resistance from editors and reviewers, we feel that showing a repeated pattern over many species is essential. We now have records on 36 species of tenthredinid sawflies (plus one pamphiliid and one cephid sawfly) that show a common pattern in their adaptive syndrome, having inherited the ability to identify and utilize the larger, more vigorous shoots (Table 5.2). Twenty-six of these species are gall inducers, representing about 20 percent of the fauna in Europe and North America (some species are common to both biogeographic realms). Even most of the species with somewhat divergent patterns of attack and survival are listed in Table 5.2 because they do adhere to the critical shoot length limitation of resources. Only the *Pontania* species that emerge very early in the spring represent a different kind of pattern, a relatively small number of species in only one genus plus the one *Nematus* species listed in Table 5.4. Even these appear to be very patchy in distribution over a landscape, and patchy relative to the distribution of willow host plants. We have yet to determine the reasons.

A third noteworthy point is the predictive power of the phylogenetic constraints hypothesis. After studying two *Euura* species we predicted that gall-inducing sawflies in general would be addicted to the rarer classes of shoots because of their common adaptive syndrome, with the general emergent properties resulting. Based on this we also predicted a biogeographic pattern in relation to plant size without any empirical hint of what the pattern might be. Whether we have the correct explanation for this pattern is difficult to assess, but we do offer a plausible mechanistic explanation for the pattern, derived from the phylogenetic constraints hypothesis.

The weight of evidence on tenthredinid sawflies we have presented so far supports the phylogenetic constraints hypothesis on this group. After our initial studies on two sawfly species, *Euura lasiolepis* and *Euura mucronata*, which provided the basis for the phylogenetic constraints hypothesis, we have found species that repeatedly show patterns consistent with the hypothesis: 11 species of *Euura*, 4 species of *Pontania*, 9 species of *Phyllocolpa*, 10 species of free-feeding sawfly, and one shoot borer in the genus *Ardis* (Table 5.2). In fact, predictions from the hypothesis proved to be more broadly applicable than we had anticipated, for we expected that free-feeding sawflies would not fit the pattern until Carr's original studies starting in 1991 (Carr 1995; Carr *et al.* 1998). Indeed, such consistency in results prompts me to venture that we have a scientific theory, macroevolutionary by nature, that accounts for macroecological patterns in the group. There are bound to be exceptions, which we need to study in more detail, and there are inevitably major divergences by groups with the same phylogenetic constraint, as we discuss later for the conifer sawflies in the family Diprionidae. The question at present, approached in the next chapter, is "How far can the Phylogenetic Constraints Hypothesis be extended in relation to other taxa with similar constraints?"

Undoubtedly, there is in these developments the basis for a paradigm shift in the way we cogitate upon and research the central issues in ecology: distribution, abundance, and population dynamics. We can move from idiosyncratic to nomothetic studies, from microecology to macroevolution, and to a macroecological theory embracing macroevolutionary mechanistic explanations for pattern, and we can enter into a phase of ecology that is fully embraced by the evolutionary synthesis. We can jettison death tables, with their inevitable results and the kinds of mentality associated with them, and seek a balanced view of bottom-up, top-down, and lateral influences on a population, recognizing that natality and mortality must both be well studied (cf. Jones *et al.* 1997; Godfray and Müller 1998). The nature of resources in relation to the needs of the exploiting population is probably more important than natural enemies in the evolution of a species because 100 percent of individuals require food and a place to live. As we will see in Chapter 8, differences in resources can select for very different adaptive syndromes even in sister families of sawflies.

6

Convergent constraints in divergent taxonomic groups

If the Phylogenetic Constraints Hypothesis applies to many species of sawflies, then it is likely to work in equivalent ways on other gall-inducing taxa. Even endophytic species, involving placement of an egg into plant tissue but without gall induction, may well evolve with equivalent constraints, adaptive syndromes, and emergent properties. These kinds of species are the subject of this chapter. We expect the species to be gall inducers or otherwise endophytic, to have life history traits consistent with the Phylogenetic Constraints Hypothesis, to respond positively and strongly to shoot length and or general plant vigor, and to be relatively uncommon or rare, patchily distributed, with latent population dynamics.

These groups offer the chance to test the Phylogenetic Constraints Hypothesis with independent taxa, and to test repeatedly. Broadening the comparative front in the search for general patterns in nature is the next logical step after moving from a focal species (Chapters 3 and 4) to its related species (Chapter 5).

Oviposition by insects into plant tissues is a relatively common trait, having evolved many times in strongly divergent phylogenetic lineages. Coupled with this behavior is the potential for the evolution of gall formation. This is because a female probes plant tissue to place the egg and uses lubricants from specific glands to ease the egg's passage through a narrow and often elongated egg canal in the ovipositor. It is therefore easy and even likely that fluids from the female will happen to stimulate cell division or cell enlargement, resulting in small swellings, which may prove to be beneficial to larvae. This is in fact the case for the gall-inducing sawflies. The female adult induces the gall with secretions even before laying the egg (cf. Fig. 3.8 where galls are formed but no egg is deposited in many cases), and the gall grows to full size in some species whether a larva is present or absent (Price and Clancy 1986b).

However, gall induction by sawflies is unusual, the more common mechanism being nymphal or larval stimulation once it hatches
from the egg. Larval mandibular gland secretions may contain agents
that stimulate hypertrophy, even initially as a plant defense. Among
the sucking insects, inserting mouthparts into plant tissue and injecting salivary fluid is a feeding mechanism prone to establish the
gall-inducing habit and several groups have evolved in this direction
(Table 6.1). Other gall-inducing taxa are listed in Table 6.1.

THE GALL-INDUCING ARTHROPODS

Feeding in galls is one of the major feeding strategies observed in insects, with many thousands of species represented. The richest family
of phytophagous insects in Britain, in one of the best-studied faunas in
the world, is the gall midges in the family Cecidomyiidae (Price 1980).
With 629 species described they vastly outnumber the more commonly
studied species of butterflies (less than 70 species) and dragonflies (less
than 50 species). The eighth-largest family in Britain is the Cynipidae,
the gall wasps so commonly found on oaks and roses. In North America
about 900 species of Cecidomyiidae are associated with plants (Gagné
1989) but many remain undescribed. Many other taxa have members
and lineages that have evolved into the galling habit, so that world-
wide there can be little doubt that tens of thousands of species exist,
although largely undocumented.

We have good reason to expect that gall inducers in general will
show patterns equivalent to the tenthredinid sawflies whether or not
they oviposit into plant tissue. Most significantly, a gall inducer causes
reprogramming of plant development, almost always involving induction during early development of a host module. Therefore, young,
rapidly developing and vigorous growth is likely to be most favorable
to the galling habit, just as in the sawflies. In addition, piercing of
plant parts is most easily accomplished in young soft tissue. For sucking insects, in which nymphs initiate gall growth, high host plant nutrient status, coupled with young modules, will be most favorable. Then,
the architecture of shrubs is likely to be more conducive to sustaining
galling populations than trees, at least for some groups, because apical dominance is absent and young sprouting shoots from the plant
base provide a more reliable resource. We can test the results of these
predictions by various methods, discussed below.

One of the most highly regarded ecological studies on plant and
gall-inducing insect interactions is by Whitham (1978, 1980). His

Table 6.1. *Gall-inducing species that show a positive response to module size or plant vigor in general[a]*

Taxonomic group/ insect species	Plant species	Locality	r^2	p	Reference
Hymenoptera: Cynipidae: Gall wasps					
1. *Andricus kollari*	*Quercus robur*	Surrey, U.K.			M. Crawley,
2. *Andricus lignicola*	*Quercus robur*	Surrey, U.K.			pers. comm.
3. *Andricus* sp.	*Quercus gambelli*	Flagstaff, Arizona	0.97	<0.01	Pires and Price 2000
4. *Diplolepis fusiformans*	*Rosa arizonica*	Flagstaff, Arizona	0.84	<0.01	Caouette and Price 1989
5. *Diplolepis spinosa*	*Rosa arizonica*	Flagstaff, Arizona	0.92	<0.01	Caouette and Price 1989
6. *Disholcaspis cinerosa*	*Quercus* sp.	Dallas, Texas			Frankie and Morgan 1984
7. *Xanthoteras politum*	*Quercus stellata*	New Jersey Pine Barrens			Washburn and Cornell 1981
Diptera: Cecidomyiidae: Gall midges					
8. *Contarinia* sp.	*Pseudotsuga menziesii*	British Columbia			Karban 1987
9. *Contarinia* sp.	*Palicourea rigida*	Brasilia, Brazil	0.89	<0.002	Vieira *et al.* 1996
10. *Rhabdophaga rosaria*	*Salix phylicifolia*	Joensuu, Finland	0.74	<0.01	Price *et al.* 1990
11. *Rhabdophaga strobiloides*	*Salix chordata*	Pellston, Michigan			Weis and Kapelinski 1984
12. *Rhabdophaga terminalis*	*Salix alba, S. fragilis*	Uppsala, Sweden			Åman 1984
13. *Cecidomyiid* sp.	*Platypodium elegans*	Belo Horizonte, Brazil	0.81	<0.01	P. W. Price, unpublished data
14. *Cecidomyiid* sp.	*Bauhinia* sp.	Isla Pirapitinga, Brazil	0.80	<0.005	W. Fernandes, unpublished data
15. *Anadiplosis* nr. *venusta*	*Machaerium angustifolium*	Belo Horizonte, Brazil	0.92	<0.01	P. W. Price, unpublished data
16. *Rhabdophaga* sp.	*Salix lutea*	Weber River, Utah			R. DeClerck-Floate, unpublished data
17. *Rhabdophaga* sp.	*Salix novae-angliae*	Tanana River, Alaska			Roininen *et al.* 1997
18. *Tokiwadiplosis matecola*	*Lithocarpus edulis*	Kagoshima, Japan			Okuda and Yukawa 2000

Table 6.1. (cont.)

Taxonomic group/ insect species	Plant species	Locality	r^2	p	Reference
Diptera: Agromyzidae: Leaf-miner flies (except for the genus Hexomyza)					
19. *Hexomyza schineri*	*Populus tremuloides*	Tanana River, Alaska	0.91	<0.01	Roininen *et al.* 1997
		Flagstaff, Arizona			Roininen *et al.* in press
20. *Hexomyza simplicoides*	*Salix chaenomeloides*	Kyoto, Japan	0.22	<0.01	Yamazaki 2001
Homoptera: Adelgidae: Pine and spruce aphids					
21. *Adelges cooleyi*	*Picea engelmanni*	San Francisco Peaks, Arizona			Fay and Whitham 1990
Homoptera: Aphididae: Plant lice or aphids					
22. *Pemphigus betae*	*Populus angustifolia*	Weber River, Utah	0.88	<0.05	Whitham 1978
23. *Pemphigus* sp.	*Populus balsamifera*	Tanana River, Alaska			Roininen *et al.* 1997
24. Wax currant aphid shoot galler	*Ribes cereum*	Flagstaff, Arizona	0.64	<0.01	Price *et al.* 1990
Homoptera: Psyllidae: Jumping plant lice					
25. Psyllid sp.	*Myrcia itambensis*	Belo Horizonte, Brazil	0.68	<0.01	Price *et al.* 1995b
Homoptera: Phylloxeridae: Phylloxerans					
26. *Daktulosphaira vitifoliae*	*Vitis arizonica*	Oak Creek, Arizona			Kimberling *et al.* 1990
Lepidoptera: Gelechiidae: Gelechiid moths					
27. Amorpha shoot galler	*Amorpha fruticosa*	Lincoln, Nebraska	0.68	<0.01	Price *et al.* 1990
28. *Amblypalpis olivierella*	*Tamarix nilotica*	Caesarea, Israel	0.96	<0.01	Price and Gerling in press

[a] r^2 is the square of the correlation coefficient which provides an estimate of the amount of variation in the data accounted for by the regression equation; p is the probability of a significant result being in error.

conclusions are similar to those advanced here, because females select the largest modules available to them and they reproduce most effectively on such modules. There is a very strong preference–performance linkage, although it is an early, first instar nymph that shows the preference; she forms the gall, and it is her performance as an adult that depends on module size. Whitham studied the aphid *Pemphigus betae* which forms galls on the narrowleaf cottonwood, *Populus angustifolia*. In very early spring, females hatch from eggs on the trunk's bark and climb up and out to where buds are just breaking, where they initiate gall formation. Although only about 0.6 mm long, females will joust for the best position on a leaf, that being the most basal part of the midrib. On large leaves all females successfully form galls, but as leaf area declines there is an increase in failure rate or aborted attempts (Fig. 6.1). In addition, as leaves become larger, the number of parthenogenetically produced progeny increases and the weight of the reproductive female increases, the stem mother (Fig. 6.2). Jousting displaces weaker females to more distal positions on the midribs and they suffer a loss in fecundity, increased rates of gall abortion, and reduced weight (Fig. 6.2).

The similarities between *Pemphigus betae* and *Euura lasiolepis* are striking. Comparing Figs. 3.6 and 6.1, we see many more galls formed on large modules than on small modules. Comparing Figs. 3.8 and 6.1, many more galls fail on smaller modules, remembering that in the low-water treatment in Fig. 3.8 willows grow poorly and shoots are generally short, and abortion would be recorded if the gall contained no egg. In addition, at the time of gall dissection for *Euura lasiolepis* late in the life cycle, very early death of larvae may be concealed by parenchymatous growth within the galls, so this would be classed as abortion also. Thus, all loss of a potential cohort, from gall formation to living second instar larvae in Fig. 3.8, would be classed as abortion. This constitutes the major difference between cohort survival on high- and low-quality shoots, just as it does in high- and low-quality leaves for *Pemphigus betae*. A difference between these species is that *Pemphigus* attacks large trees rather than young trees. Perhaps a constraint exists because overwintering eggs are placed in fissured bark on large trunks, lacking on small trees.

Many other gall-inducing insects show equivalent patterns in relation to shoot length and plant vigor (Table 6.1). The general trend in these examples is for more galls to be formed on larger modules, whether the gall inducers form stem or leaf galls, because leaf size and shoot length are positively correlated. The 28 species listed

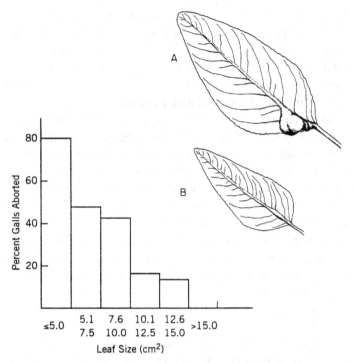

Fig. 6.1. The relationship between leaf area and percentage abortion of gall-inducing attempts by the aphid *Pemphigus betae* on its host plant, *Populus angustifolia*, when only one female per leaf attempts to initiate a gall. (A) Successful development of a gall. (From Whitham 1980) (B) An aborted gall. (From Whitham 1978.) ((A) reproduced from Whitham, T. G. (1980) The theory of habitat selection, *Am. Nat.* 115: 449–466, University of Chicago Press.)

lend strong support to the phylogenetic constraints hypothesis developed on gall-inducing sawflies. Although each taxonomic group has evolved the gall-inducing habit independently, there is impressive convergence in responses to plant architecture repeatedly: in gall wasps, gall midges, the one genus of leaf-mining agromyzids that forms galls (nos. 19 and 20 in Table 6.1), adelgids, aphids, psyllids, and moths. Seven independent lineages have evolved to respond positively to host plant shoot length consistent with the 26 species of gall-inducing sawflies listed in Table 5.2. Thus, the general pattern in gall inducers is clear.

    We have found a small number of gall inducers that do not fit this general pattern, as in the *Pontania* species with very early emergence relative to the phenology of the host plant shoot growth. Deviation

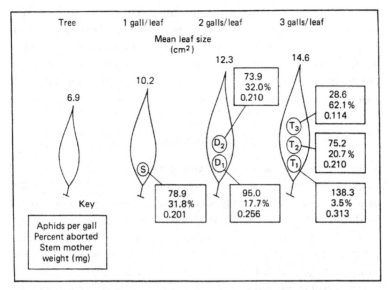

Fig. 6.2. The effect of gall position on female fitness of *Pemphigus betae* females on small to large leaves and when multiple galls are formed on a leaf. Boxes contain information on (top) number of aphids per gall; (middle) percent aborted; (bottom) weight of the stem mother, which is the fully grown female that has initiated the gall. (From Whitham, T. G. (1980) The theory of habitat selection, Fig. 6, *Am. Nat.* 115: 449–466, University of Chicago Press.)

from the pattern also appears to occur in the tropics, where plant architecture is frequently complicated by asynchrony within the canopy of individual woody plants: some branches will support developing shoots while others will be dormant. Thus, a more complicated picture develops and more detailed studies are needed to record attack by gallers on the developing shoots. This we did in the Mediterranean climate of Israel, finding that *Tamarix* growth is complex but, where this was understood, we could detect a positive response of the spindle-gall moth larvae, *Amblypalpis olivierella*, to shoot length and vigorous modules (Price and Gerling in press; no. 28 in Table 6.1).

Getting around the complex phenology problem in warm temperate and tropical latitudes is easier in vegetation in which fire is frequent, as in the cerrado of Brazil. In this savanna-like growth of fire-tolerant species a fire sets all plants equal in phenology, for resprouting occurs rapidly after fire. Then we could find repeated examples of both gallers and nongallers responding positively to plant module size and vigor. One example is a *Contarinia* sp. (no. 9 in Table 6.1) on

Palicourea rigida at the Fazenda Agua Limpa field station of the University of Brasilia (Vieira *et al.* 1996). When complex plant phenology is simplified by fire, then this provides an ideal opportunity for studying the response of herbivores to vigorous regrowth after the fire, and growth in adjacent unburned sites (cf. Prada *et al.* 1995; Seyffarth *et al.* 1996; Vieira *et al.* 1996).

BIOGEOGRAPHY OF GALL-INDUCING INSECTS

If the architecture of shrubs is conducive to the galling habit, and frequent fires produce flushes of new and vigorous growth, then it is likely that gall species richness is high in climatic zones dominated by shrubs and dry seasons in which fire is common. These warm temperate, Mediterranean-type climates are extensive world-wide because a lot of the tropics at higher elevations supports vegetation somewhat convergent with the fire-adapted true Mediterranean vegetation: deserts, scrublands, and savannas. Caatinga, campina, cerrado, chapparal, matorral, kwongan, fynbos, and maquis all contain a rich woody flora, scleromorphic and fire-adapted. Might these vegetations be centers of diversification of gall-inducing species?

In our various peregrinations we sampled local gall species richness. Coupled with friends who added sites, we censused more than 280 sites from around the world, with centers of sampling noted in Fig. 6.3 (Price *et al.* 1998b). All continents except Antarctica were represented and all biogeographic realms except the Oriental. Correcting all altitudes of samples to their equivalent latitude, by Merriam's conversion of 4 degrees latitude for every 305 m (1000 feet) increase in elevation, we found a remarkably strong pattern of local species richness in warm temperate latitudes and their tropical equivalents (Price *et al.* 1998b) (Fig. 6.4). Where galling species number in local samples exceeded 12, vegetation types were exclusively scleromorphic, occurring around the world: Australia, Sonoran Desert of Arizona, Israel, campina and cerrado in Brazil, and fynbos in South Africa.

There is, indeed, a very rich gall-inducing fauna on drought- and fire-adapted woody plants, many of which are shrubs. A mechanistic hypothesis has been developed by Fernandes and Price (1991) based on Wilson Fernandes's detailed studies in Arizona and in the cerrado of Minas Gerais State, in Brazil. We would certainly advocate the inclusion of fire as an additional favorable characteristic of scleromorphic vegetation, promoting the quality of life for gallers, resulting in a revised hypothesis, as in Fig. 6.5 (see also Mendonça 2001).

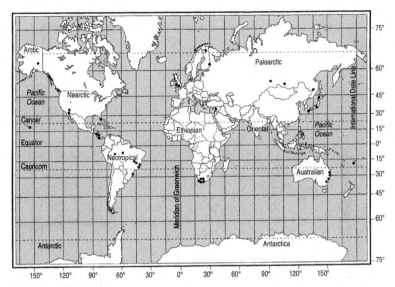

Fig. 6.3. Map of the world showing distribution of clusters of local sites sampled to estimate local species richness of gall-inducing insects. The map uses the Mercator projection. (From Price, P. W., G. W. Fernandes, A. C. F. Lara, J. Brawn, H. Barrios, M. G. Wright, S. P. Ribeiro, and N. Rothcliff (1998) Global patterns in local number of insect galling species, Fig. 1, J. *Biogeog.* 25: 581–591, Blackwell Science, Oxford.)

An interesting departure from the general pattern of females showing a preference for the longest shoots available is seen in comparisons among xeric and mesic sites within the same locality and latitude. For example, we compared xeric sites away from water with riparian sites, and with the same host plant species present in each. Although plants in drier sites grew less vigorously than their counterparts in riparian vegetation, galls were more abundant in the drier sites, showing a response to the hotter, more common habitat of scleromorphic plant hosts (Fernandes and Price 1988). Nevertheless, galls on modules within plant individuals were concentrated on the more vigorous shoots. A similar pattern has been found for shoot borers in the genus *Dioryctria*, which attacks an inherently vigorous class of trees, but is most abundant on stressed sites (Ruel and Whitham in press).

Mention of shoot borers with similar patterns of plant exploitation to gallers raises the question on how general the similarities are. As stated earlier in the chapter, endophytic insects may well have phylogenetic constraints in common. These and other kinds of phytophagous arthropods are considered next.

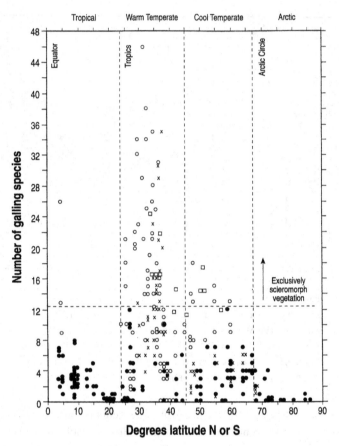

Fig. 6.4. Number of gall-inducing species in samples north or south of the equator in relation to latitude when all samples are plotted as if at sea level, having been corrected for altitude with Merriam's correction. Note the strong peak in richness in warm temperate regions or their altitudinal equivalents. Open circles represent samples on scleromorphic vegetation. Closed circles are samples from mesic sites on nonsclero-morphic vegetation. x is for samples on scleromorphic vegetation in relatively mesic sites such as riparian habitats. Squares indicate samples from fynbos vegetation in South Africa which acted as an independent test of the pattern. The open circles close to the equator show samples from campina vegetation, found as small islands of scleromorphic vegetation on the white sands along the Rio Negro, Amazonia. (From Price, P. W., G. W. Fernandes, A. C. F. Lara, J. Brawn, H. Barrios, M. G. Wright, S. P. Ribeiro, and N. Rothcliff (1998) Global patterns in local number of insect galling species, Fig. 2, J. Biogeog. 25: 581–591, Blackwell Science, Oxford.)

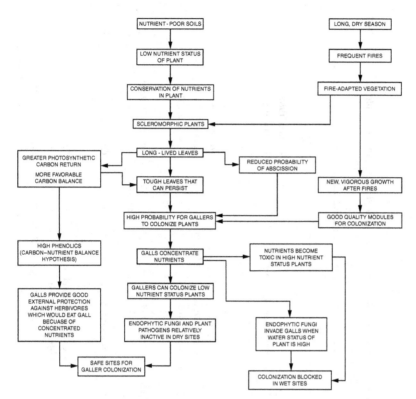

Fig. 6.5. The hypothesis to account mechanistically for the high local richness of gall-inducing insect species, based on the evolution of scleromorph vegetation in climatic zones with a long dry season. The original hypothesis was developed by Fernandes and Price (1991) and is modified here to include the favorable effect of fire on plant vigor and resources for gall-inducing insects. (From Fernandes, G. W. and P. W. Price (1991) Comparison of tropical and temperate galling species, Fig. 5.16, in P. W. Price, T. M. Lewinsohn, G. W. Fernandez, and W. W. Benson (eds.) *Plant–animal interactions*, © 1991, reprinted by permission of John Wiley & Sons, Inc.)

NON-GALLING ARTHROPODS

There is no shortage of examples of herbivores utilizing vigorous plants or plant parts. Common knowledge includes recognition of many species attacking tender parts of garden and greenhouse plants, from mites and aphids to shoot tip moths and other borers. I include just a few examples from the literature and personal contacts that illustrate the pattern in Table 6.2.

Table 6.2. *Arthropods other than gall inducers that show a positive response to large modules of host plants, or young plants, or high host plant metabolic activity, in addition to sawfly species listed in Table 5.2*

Taxonomic group/arthropod species	Plant species	Locality	Reference
Acarina: Mites			
1. *Oligonychus milleri*	*Pinus caribaea*	Jamaica	Karban 1987
2. *Oligonychus submudus*	*Pinus radiata*	Monterey, California	Karban 1987
3. *Oligonychus ununguis*	*Picea*	Finland	Karban 1987
Homoptera: Aphididae: Aphids			
4. *Kallistaphidus betulicola*	*Betula pubescens*	Skipwith Common, U.K.	Fowler 1985
5. *Mindarus abietinus*	*Abies concolor*	Sierra Nevada, California	Karban 1987
6. *Symydobius oblongus*	*Betula pubescens*	Kevo, Finland	Price *et al.* 1990
Homoptera: Aleyrodidae: Whiteflies			
7. *Bemisia argentifolii*	*Lycopersicon esculentum*	Growth chamber	Inbar *et al.* 2001
Homoptera: Cercopidae: Froghoppers			
8. *Aphrophora pectoralis*	*Salix* sp.	Sapporo, Japan	Craig and Ohgushi in press
Homoptera: Membracidae: Tree hoppers			
9. *Oxyrachis tarandus*	*Tamarix nilotica*	Caesarea, Israel	Price and Carr 2000
10. *Encenopa* sp.	*Caesalpinia* sp.	Brasilia, Brazil	Price and Carr 2000
11. *Leioscyta* sp.	*Myrcia* sp.	Serra do Japi, Campinas, Brazil	Price and Carr 2000
12. *Heteronotus formicoidea*	*Acacia polyphylla*	Serra do Japi, Campinas, Brazil	Price and Carr 2000

	Eupatorium maximiliani	Serra do Japi, Campinas, Brazil	Price *et al.* 1995b
Homoptera: Cicadellidae: Leaf hoppers			
13. *Aulacizes* sp.			
Homoptera: Delphacidae: Plant hoppers			
14. *Prokelisia marginata* + 5 other species	*Spartina alterniflora*	Mullica River, New Jersey	Denno *et al.* in press
Coleoptera: Beetles			
15. *Chrysomela confluens*	*Populus angustifolia*	Weber River, Utah	Kearsley and Whitham 1989
16. *Cylindrocopturus eatoni*	*Pinus ponderosa* etc.	Northren California	Eaton 1942
17. *Hylobius pales*	*Pinus*	Ontario, Canada	Finnegan 1959
18. *Hylobius radicis*	*Pinus*	Ontario, Canada	Finnegan 1962
19. *Merhynchites bicolor*	*Rosa arizonica*	Flagstaff, Arizona	P. W. Price, unpublished data
20. *Pissodes approximatus*	*Pinus*	Ontario, Canada	W. L. Baker 1972
21. *Pissodes strobi*	*Pinus strobus*	U.S.A. and Canada	Furniss and Carolin 1977
22. *Pissodes terminalis*	*Pinus*	Western U.S.A. and Canada	Furniss and Carolin, 1977
23. *Rhynchites betulae*	*Betula pubescens*	Joensuu, Finland	Price *et al.* 1990
24. *Taphrocerus schaefferi*	*Cyperus esculentus*	Montgomery Co., Virginia	Story *et al.* 1979
25. *Tomicus piniperda*	*Pinus sylvestris*	Simonstorp, Sweden	Långström 1980
26. Weevils	*Aspilia foliaceae*	Brasilia, Brazil	Prada *et al.* 1995
27. *Acrocercops brongniadella*	*Quercus robur*	Surrey, U.K.	M. Crawley, pers. comm.
28. *Cycnotrachelus roelofsi*	*Styrax japonicus*	Fukuoka Prefecture, Kyushu, Japan	Tokuda *et al.* 2001

Table 6.2. (cont.)

Taxonomic group/arthropod species	Plant species	Locality	Reference
Lepidoptera: Moths			
29. *Dioryctria albovitella*	*Pinus edulis*	Sunset Crater, Arizona	Mopper and Whitham, 1986
30. *Dioryctria ponderosae*	*Pinus ponderosa*	Western U.S.A.	Furniss and Carolin 1977
31. *Eriocrania subpurpurella*	*Quercus robur*	Surrey, U.K.	Michael Crawley, personal communication
32. *Eucosma gloriola*	*Pinus banksiana*	Lake States, U.S.A.	W. L. Baker 1972
33. *Eucosma sonomana*	*Pinus banksiana*	Lake States, U.S.A.	W. L. Baker 1972
34. *Gypsonoma haimbachiana*	*Populus deltoides*	Eastern U.S.A.	W. L. Baker 1972
35. *Heliothis zea*	*Lycopersicon esculentum*	Growth chamber	Inbar *et al.* 2001
36. *Lithocolletis salicifoliella*	*Populus tremuloides*	Sault Ste. Marie, Ontario	Martin 1956
37. *Mitoura gryneus*	*Juniperus* spp.	Eastern U.S.A.	Karban 1987
38. *Petrova albicapitana*	*Pinus banksiana*	Northern U.S.A. and Canada	W. L. Baker 1972
39. *Phyllonorycter strigulatella*	*Alnus incana*	Kola Peninsula, Russia	Kozlov in press
40. *Phyllonorycter* sp.	*Salix lasiolepis*	Flagstaff, Arizona	Preszler and Price 1995
41. *Rhyacionia buoliana*	*Pinus*	Eastern U.S.A.	W. L. Baker 1972
42. *Rhyacionia frustrana*	*Pinus*	Eastern U.S.A.	W. L. Baker 1972
43. *Rhyacionia neomexicana*	*Pinus ponderosa*	Flagstaff, Arizona	Spiegel and Price 1996
44. *Satyrium caryaevorus*	*Carya* spp.	Eastern U.S.A.	Karban 1987
45. *Satyrium edwardsii*	*Quercus* spp.	Eastern U.S.A.	Karban 1987

46. *Satyrium falacer* (= *calanus*)	*Quercus* spp.	Eastern U.S.A.	Karban 1987
47. Leaf rollers	*Ouratea hexasperma*	Brasilia, Brazil	Seyffarth *et al.* 1996
48. *Cactoblastis cactorum*	*Opuntia stricta*	Hunter Valley, New South Wales, Australia	Stange *et al.* 1995
49. *Helicoverpa armigera*	About 200 plant species	Laboratory	Rasch and Rembold 1994

Diptera: Flies

50. *Agromyza frontella*	*Medicago sativa*	Laboratory	Drolet and McNeil 1984
51. *Liriomyza trifolii*	*Lycopersicon esculentum*	Growth chamber	Inbar *et al.* 2001
52. *Phytobia betulae*	*Betula pendula* and *pubescens*	Valtatie, Finland	Ylioja *et al.* 1999
53. *Orellia occidentalis*	*Cirsium wheeleri*	Flagstaff, Arizona	Fondriest and Price 1996
54. Tephritids	*Aspilia foliacea*	Brasilia, Brazil	Prada *et al.* 1995

General:

| 55. Psyllids, leaf gallers, leaf miners | *Betula pubescens* and *pendula* | Umeå, Sweden | Danell and Huss-Danell 1985 |
| 56. Mirids, membracids, cicadellids and thrips | *Larrea tridentata* | Chihuahuan Desert, New Mexico | Lightfoot and Whitford 1987 |

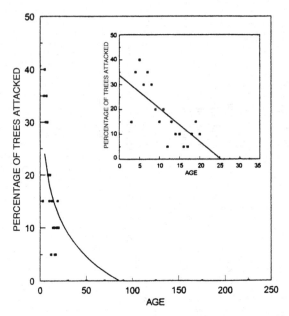

Fig. 6.6. The percentage of ponderosa pine trees attacked by the south-western pine tip moth, *Rhyacionia neomexicana*, relative to the age of the trees, around Flagstaff, Arizona. The inset covers only ages 1–25 years, indicating no attacks on trees older than 20 years. The pattern in this figure is comparable to that for the sawfly *Euura amerinae* in Fig. 5.3. (From Spiegel and Price 1996.)

The list includes very rare insects over a landscape such as the rose curculio, *Merhynchites bicolor* (no. 19 in Table 6.2), with only one tiny population known to me on a few square meters of its local host plant, *Rosa arizonica*. Others on the list are considered to be serious forest pests, such as the pine tip moths (*Rhyacionia*) and pine shoot moths (*Eucosma*), but most importantly in plantation settings. In a natural forest, young plants would be patchily distributed, and hard to find. Our local ponderosa pine forests, around Flagstaff, Arizona, are certainly not of primaeval character but neither are they managed intensively. In relation to the southwestern pine tip moth, *Rhyacionia neomexicana*, stands of young ponderosa pine suitable for attack are very patchy. Some are colonized while some are not. No attacks were found on trees older than 20 years and taller than 4 m, although such trees were rare in a landscape dominated by much older stands, ranging up to 300 years in the study (Spiegel and Price 1996). Frequency of trees attacked declined rapidly with tree age and the most vigorous leading shoots were attacked most frequently (Fig. 6.6). Note the similarity of this pattern with that shown in Fig. 5.5 for *Euura mucronata*. Females clearly select oviposition sites with precision close to the largest buds on young trees, and larvae hatch and enter these buds. The proximity of oviposition site and larval feeding site is such that a high preference–performance linkage is likely to evolve. Evidence suggests that trees older than 20 years become resistant to attack, perhaps involving ontogenetic aging. Similarities to galling sawflies and other gall-inducing species are evident.

Many of the species listed in Table 6.2 are endophagous as larvae: *Hylobius* and *Pissodes* weevils; *Taphrocerus*, *Eriocrania*, *Lithocolletis*, *Agromyza*, and *Phyllonorycter* leaf miners; *Dioryctria*, *Petrova*, and *Rhyacionia* shoot and pitch moths; *Orellia* and other tephritid flower head dwellers as well as weevils; *Rhynchites*, a leaf roller; and the amazing *Phytobia betulae*, which mines down the cambium through the length of tall birch trees. Free-living species include the mites, aphids, the cercopid *Aphrophora pectoralis* or awafuki in Japan, and membracids in the genera *Oxyrachis*, *Encenopa*, *Leioscyta*, and *Heteronotus*. Adult feeding on vigorous shoots is observed in *Tomicus piniperda*, the pine shoot beetle, and *Merhynchites bicolor*, the rose curculio, with larvae of both feeding in concealed locations.

The membracids illustrate an excellent example of convergence of phylogenetic constraints in disparate taxonomic groups, and we have dealt with the similarities of membracid tree hoppers and common sawflies at length (Price and Carr 2000). A plant-piercing ovipositor is a plesiomorphic trait in membracids as it is in tenthredinids. It is used for slitting stems for placement of eggs and vigorous plants and shoot modules are exploited. Although nymphs are exophytic, the oviposition constraint prevails in the evolution of an adaptive syndrome and the consequent emergent properties equivalent to those in the *Euura* and other sawflies. Membracids in the tropics are exceedingly patchy in distribution and low in abundance. They are most readily located in disturbed areas where plants are cut back along trails and roads and in horticultural settings. Indeed, more than once I have ventured forth to collect data on membracids in the tropics, knowing of a nearby population along a road or outside a hotel, only to find gardeners hacking back the growth and dispersing the adults!

Obviously many exceptions to these patterns exist. The challenge is to find other categories and other patterns into which different kinds of species can be placed. Karban (1987) cited Art Shapiro as the source of a personal communication for the skippers in the genera *Satyrium* and *Mitoura* attacking young plants, but at a lecture I gave recently at

Davis, California, Art was equally willing to cite many examples among the butterflies that do not fit the pattern. The Plant Stress Hypothesis by White (1969, 1974) gained many adherents, so much so that I was obliged to counter with the Plant Vigor Hypothesis (Price 1991c) based on studies of common sawflies and many of the species in Tables 6.1 and 6.2. White's (1984, 1993) subsequent emphasis on nitrogen limitation in plants brought our hypotheses much more into unison because White noted that nitrogen was most available during flushes of foliage and during senescence as nutrients moved out of leaf modules. Hence his categorization of flush feeders and senescence feeders (White 1993).

Karban (1987) proposed his Induced Defense Hypothesis to account for attack on young plants by phytophagous arthropods such as mites, while older plants were not attacked. He argued that if herbivores induced defense in young plants, these would subsequently become resistant, creating the pattern frequently observed. However, this hypothesis was not supported in his own tests (Karban 1990) and was not resurrected by Karban and Baldwin (1997). In our own studies of *Euura* species and shoot moths, induced defenses appear to play no role in modifying patterns of attack, leaving the most likely explanation to be dependent on plant aging and vigor.

There is a growing interest in detecting patterns of herbivore attack in relation to the gradient from plant stress to plant vigor. This is very encouraging but much more needs to be explored using experimental approaches such as those by English-Loeb (1989, 1990), offering a rare example of how a species responds over much of the gradient. The approaches by Larsson (1989) and Koricheva et al. (1998) are encouraging, which search for pattern in response to the host plant stress to vigor gradient, according to feeding type of the arthropod herbivore. And the review by Waring and Cobb (1992) alerts us to the problems with generalizations, such as by Larsson (1989), and even the lack of concordance among experiments and field observations.

The main point of this chapter has been to show how herbivores with similar constraints to the focal species and group discussed in Chapters 3, 4, and 5 show similar adaptive syndromes and emergent properties. The strongest constraint is oviposition into plant tissue using a plant-piercing ovipositor. Even though the taxa discussed in this chapter are distantly related to the sawflies, having diverged from a common stock many millions of years ago, they are all constrained by similar plesiomorphic traits, with resultant adaptive syndromes and emergent properties closely comparable. There is pattern in nature on a broad scale, with mechanistic understanding well developed.

We now have evidence for broad patterns in nature. In addition to the 38 species of sawfly (Table 5.2) that show responses and distributions consistent with the phylogenetic constraints hypothesis, we can add 28 species of gall-inducing insects outside the Tenthredinidae (Table 6.1) and 56+ species of nongalling insects (Table 6.2). This is weighty evidence consistent with an hypothesis, so seldom assembled on any hypothesis on plant and animal interactions. The total number of species adds up to over 120, showing consistency within a taxon, the Tenthredinidae, and similarity in patterns across many taxa, many only remotely related phylogenetically to the sawflies. In all these cases, six orders and over 20 families of arthropods are represented.

Patterns of distribution and abundance over a landscape, so far as we know them, are consistent with the hypothesis. Locally, small and patchy populations exist over a landscape, depending very much, or completely, on the quality of the host plant population. In many cases where we have extensive field experience, only one or very few sites are known in which numbers of herbivores of a particular species are high enough on which to collect adequate samples.

An encouraging aspect of the pattern is consistency of observations in temperate and tropical latitudes and vegetation types. Constraints appear to transcend large differences in climate and host plants. Patterns are defined by the close association between plant and herbivore dictated by similar phylogenetic constraints. Such constraints are so common because laying into plant tissue, or laying near large plant modules, are clearly highly adaptive aspects of maternal care.

Another quality of these patterns is the evidence derived from many kinds of plants. Conifers and angiosperms are both well represented even though their architectures are so disparate. Herbs, shrubs, and trees are all represented as host plants, as are early, mid, and late successional plant species. And plants in many biogeographic realms and their insect herbivores fit the patterns.

The phylogenetic constraints hypothesis appears to be remarkably robust in its application to a wide range of conditions, host plant types, and arthropod groups. The common theme is the intimate relationship between plant and herbivore, so much so that the insects can be regarded as parasites of plants (Price 1980), coupled with the common requirement for vigorous plants or plant modules as advocated in the Plant Vigor Hypothesis (Price 1991c). As MacArthur wrote in 1972 (1972a, p. 77), "The concept of pattern or regularity is central to science. Pattern implies some sort of repetition, and in nature it is usually an imperfect repetition. The existence of the repetition means some prediction is possible – having witnessed an event once, we can partially predict its future course when it repeats itself." We certainly can predict a lot for species with constraints similar to those species so far discussed. We can also predict high local richness of these kinds of species in sclero-morphic vegetation types, and low local richness in the wet tropics and north temperate latitudes. And we can predict patchy distributions and relatively low abundance for species conforming to similar phylogenetic constraints.

Exceptions to the patterns discussed in Chapters 3–6 are numer-ous and in need of exploration. Here we enter the pluralistic nature of the Phylogenetic Constraints Hypothesis, for with fundamentally differ-ent constraints we should expect basic differences to appear as adaptive syndromes and emergent properties. Again, MacArthur (1972b) foresaw the need for pluralism in the development of ecological theory. "All suc-cessful theories, for instance in physics, have initial conditions; with different initial conditions, different things will happen. But I think initial conditions and their classification in ecology will prove to have vastly more effect on outcomes than they do in physics." In the next chapter species with very different constraints will be discussed.

The phylogenetic constraint has not been identified explicitly for all species listed in Tables 6.1 and 6.2, involving species that oviposit on, rather than into, plants. Detailed studies are needed on such species. The point to remember, however, is that included in the adaptive syn-drome is the dependence on vigorous host plant modules, which sets the stage for emergent properties equivalent to those found for en-dophagous insect herbivores.

7

Divergent constraints and emergent properties

With so many species restricted by oviposition behaviors that result in a low carrying capacity in the environment, how is it that many
species have escaped such limitation? Why are we plagued with so many
pests in agriculture, horticulture, and forestry? One easy answer is that
humans desire high-yielding, vigorous growth on their plants and ex-
ert considerable effort in growing plants ideal for insect herbivores. Big
stems, leaves, buds, shoots, flowers, and fruits developed on nutritious
soils with a reliable water supply result from good husbandry and suit
the needs of humans and other herbivores alike. Any specialist insect
herbivore requiring large, rapidly developing plant modules for oviposition and larval food will discover a cornucopia of such parts in an
agricultural field or a managed forest, especially where monocultures
are prevalent. Thus, many species that we would predict to be limited by
resources find a bonanza of high-quality modules because of our expert
husbandry. In a wild landscape, meager growth and spotty distribution
of host plants would set the carrying capacity for insect herbivores
orders of magnitude lower than in managed croplands, gardens, and
forests.

But, there are many pests that have been outbreak species before
landscapes were severely modified by humans. For example, growth-
ring analysis of host trees in eastern Canada indicated episodes of
heavy spruce budworm (*Choristoneura fumiferana*) defoliation back into
the 1700s (Blais 1958, 1962, 1965), and sediments from Maine, bear
abundant microlepidopteran head capsule fossils back to more than
10 000 years before the present (Anderson *et al.* 1986). Identity to species
is uncertain but the head capsules belong to species in the Tortricidae,
they are probably in the genus *Choristoneura*, and the most likely
candidate is the spruce budworm. Tree-ring analyses on Douglas fir,
Pseudotsuga menziesii, host of the western spruce budworm, *Choristoneura*

occidentalis, also show regional-scale outbreaks of budworm during the last 300 years in Oregon and New Mexico (Swetnam and Betancourt 1998) and more than 400 years in New Mexico, USA (Swetnam and Lynch 1993). With considerable evident periodicity in precipitation and fires for more than 2000 years in the southwest United States, we should not be surprised if budworm have been erupting over a similar period (cf. Grissino-Mayer 1996; Swetnam and Betancourt 1998).

We can examine the spruce budworms as archetypal cases of outbreak species, for they share important traits with other pest species in forests. With budworms as the introduction to outbreak species, I will then proceed with a broader search for traits in common among outbreak forest lepidopterans and other taxa.

THE SPRUCE BUDWORM LIFE CYCLE AND BEHAVIOR

Females of the spruce budworm emerge in late July and early August in New Brunswick, Canada, when foliage of its major host balsam fir, *Abies balsamea*, is mature (Miller 1963). Lacking a plant-piercing ovipositor, as in the majority of Lepidoptera, eggs are laid on foliage, clustered together typically in groups of 20. Total fecundity is about 200 eggs per female. Larvae emerge after 8–12 days; they do not feed, but move upwards toward the light. If disturbed, larvae drop on a silken thread and are liable to disperse on this thread away from the natal tree to other trees in a stand or beyond. Many larvae disperse to nonhost plants and die, commonly accounting for more than 50 percent mortality at this stage (Miller 1958; cf. Tables 2.1 and 2.2). Those that settle on a host plant spin a hibernaculum, molt to the second instar, and remain until the following May when larvae emerge. They move to the tips of branches and readily disperse on silken threads again. On a suitable host they then start feeding by entering staminate cones, or by mining in 1- and 2-year-old needles, or by mining into expanding vegetative buds. By late May or early June larvae molt to the third instar and feed in the cluster of needles provided by newly opened vegetative buds, webbing needles together into a feeding shelter. The remaining instars feed until about mid July when pupation occurs in the feeding site of the sixth instar larva. Adults emerge after about 8–12 days (see Miller 1963 for details).

Oviposition on mature foliage, with nonfeeding first instar larvae, is a phylogenetically primitive trait in the genus *Choristoneura* shared by all species with reported life histories (cf. Baker 1972; Furniss and Carolin 1977; Nealis and Lomic 1994). Here we observe a suite of traits that are phylogenetically primitive. Females lack a plant-piercing

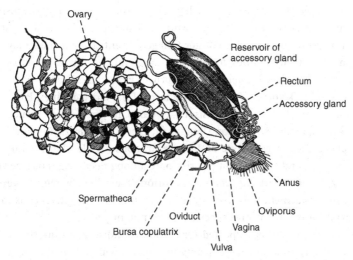

Fig. 7.1. The female reproductive system of a moth, the eastern tent caterpillar, *Malacosoma americanum* (from Snodgrass 1935). Eggs are formed in the ovary and pass down the oviduct to the oviporus, the cushion-like egg exit. During copulation sperm entered the bursa copulatrix via the vulva, and then it moved into the spermatheca for storage until oviposition. The accessory glands secrete a substance that is stored in the reservoir of the accessory glands and, as a clutch of about 200 eggs is laid on twigs and branches of host trees, the accessory gland secretion coats the clutch with a thick, frothy protective layer which hardens in air. Compare the oviporus of moths with the ovipositor of sawflies in Fig. 3.4. (Reprinted from R. E. Snodgrass (1993) *Principles of insect morphology*. Copyright © 1993 Ellen Burden and Ruth Roach. Used by permission of the publisher, Cornell University Press.)

ovipositor. As in the majority of Lepidoptera, there is a single oviporus "serving only for the discharge of eggs" with no piercing appendages (Snodgrass 1935, p. 563). Eggs are laid *onto* plant surfaces (Fig. 7.1). Females oviposit on mature foliage, and larvae have a high probability of dispersing on a silken thread twice, as a first instar larva in late summer and as a second instar larva in the spring, both before any feeding occurs. Females cannot choose suitable feeding sites for larvae because they oviposit on old foliage 9 months before larvae start feeding. Then larvae feed mostly on staminate cones, or buds or very new foliage, and this after a high probability of dispersal away from the natal site. There is certainly no ovipositional preference and larval performance linkage of the kind we saw for gall-inducing sawflies (Chapter 3), and we should not be surprised to find females showing very little preference of any kind in terms of oviposition sites.

Indeed, females are rather indiscriminate in their oviposition habits, even to the point of using brown paper bags as a substrate in oviposition tests with conifer foliage present. They will also lay eggs on almost any conifer foliage in trials, and are found in nature on balsam fir, *Abies balsamea*, white spruce, *Picea glauca*, red spruce, *Picea rubens*, black spruce, *Picea mariana*, larch, *Larix laricina*, pines, *Pinus* spp., and hemlock, *Tsuga canadensis* (Baker 1972). In the western spruce budworm, females oviposit on Douglas fir, *Pseudotsuga menziesii*, Engelmann spruce, *Picea engelmannii*, and white fir, *Abies concolor*, both young and old trees, and they do not discriminate among needle ages or tree vigor (Leyva *et al.* 2000). Aspects of ovipositional decisions by female western spruce budworm were also consistent with the phylogenetic constraints hypothesis for eruptive species (Leyva *et al.* in press).

As should be expected for larvae of undiscerning mothers, they are also general in their capacity to utilize a wide range of food quality sufficiently well to mature and produce viable offspring. For the spruce budworm, Blais (1952) found that fifth and sixth instar larvae can mature on old foliage, although reduced pupal size and fecundity result. In the western spruce budworm there were no significant differences in female pupal weight, which correlates with fecundity, when fourth instar larvae were fed on Douglas fir, Engelmann spruce, or white fir, 1-year-old or 4-year-old needles of Douglas fir, young trees or old trees of Douglas fir, or long vigorous or short less vigorous shoots of Douglas fir (Leyva *et al.* 2000).

One may ponder why the female does not oviposit on buds, or male cones or even into them, so that larvae could be protected and not have to forage for food. Why hasn't a plant-piercing ovipositor evolved in tortricid moths or the genus *Choristoneura*? Females are constrained by their simple oviporus to lay eggs on foliage, not into foliage, and this constraint acts as part of the phylogenetic constraint for the genus. Another feature in the life cycle is also significant: larval dispersal on a silken thread. There is no ovipositional preference and larval performance linkage and no positive feedback from larval survival back to female oviposition behavior. As we saw for the winter moth in Chapter 2, there is strong selection in budworms favoring feeding on very young foliage by early instars. The solution is very different in budworms compared to winter moth, but the result is essentially the same: young larvae feed on the youngest possible foliage. Without this evolutionary explanation for the need of synchrony between young foliage and young larvae, the life cycle of the budworms, and many other tortricids, would seem totally enigmatic.

APPLYING THE PHYLOGENETIC CONSTRAINTS HYPOTHESIS

The oviporus of the Lepidoptera (Fig. 7.1) is a plesiomorphic trait that defines the opportunities and constraints typical of the taxon. Snodgrass (1935) was strict in distinguishing the oviporus of the Lepidoptera and the ovipositor of many other taxa, such as in the Hymenoptera. The oviporus he defined as "The posterior opening of the vagina in most Lepidoptera, serving only for the discharge of the eggs when there are two genital apertures" (cf. Fig. 7.1, vulva and vagina; Snodgrass 1935, p. 622). By contrast, the ovipositor was defined as "The egg-laying organ formed of the gonopods of the eighth and ninth abdominal segments; or also, in a functional sense, the egg-laying tube of some insects formed of the protractile terminal segments of the abdomen" (Snodgrass 1935, p. 622). In other words, lepidopterans in general and the spruce budworm in particular lack an ovipositor.

Lack of an ovipositor might be regarded as a general release from the phylogenetic constraints associated with a plant-piercing ovipositor as in the sawflies. Eggs can be deposited on leaves, tree trunks, twigs, or rocks, or dropped during flight. However, the oviporus has its own limitations. Its chemoreceptors receive only limited information on the status of the host plant, whereas an inserted ovipositor can sense the internal state of the plant (see Chapter 3 for discussion of evidence). Even on a leaf the cuticle offers few cues to the sensors on the oviporus. And the oviporus is generally suited to laying on almost any substrate such that the female need not be exact in egg placement, letting the larva perform the final decision on where to feed. The oviporus is permissive on egg placement, opening opportunities for rather sloppy oviposition tactics and the more general use of substrates, including more than one host plant species.

Nevertheless, the oviporus acts as a phylogenetic constraint in the Lepidoptera. It limits the amount of information a female receives from the host plant because only surface stimuli are available. Therefore, discrimination among host plant species is reduced, variation in host plant quality relative to module or plant vigor or age is less likely to be perceived, and females are likely to be more catholic in where they lay their eggs.

Therefore, in the budworm we observe an evolutionary scenario with emergent properties at the opposite end of the spectrum from the *Euura* sawflies discussed in Chapters 3 and 4. Female budworms are not specific in their oviposition behavior, larvae are general in what they can mature on in relation to host species, vigor, and age, and there

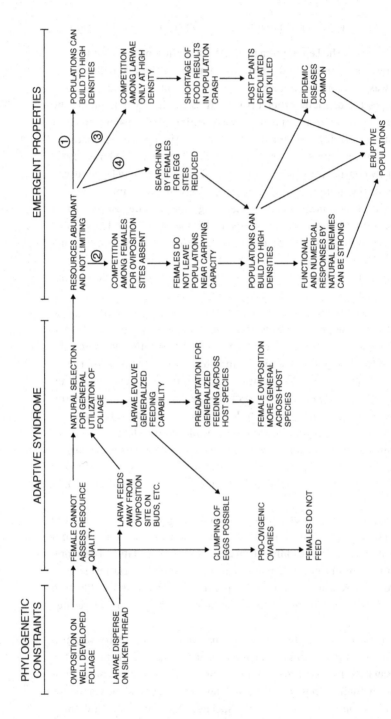

Fig. 7.2. The flow of influences in the spruce budworm, *Choristoneura fumiferana*, from the phylogenetic constraints, to the adaptive syndrome, to emergent properties. (Slightly modified from Price 1990b.)

is no preference–performance linkage. The Phylogenetic Constraints Hypothesis then follows the course illustrated in Fig. 7.2, based on Price *et al.* (1990) and Price (1990a).

Oviposition by the budworm on mature foliage, on which larvae cannot feed in the spring, and 9 months before larvae start feeding, is a phylogenetic constraint because females cannot assess resource quality for larvae, and are therefore unable to select high-quality ovipositional sites (Fig. 7.2). This divorce between female behavior and larval feeding is compounded by larvae commonly dispersing on silken threads from the natal site, with two episodes involved in fall and spring.

The adaptive syndrome then follows logically from these constraints. Since females cannot assess resource quality for larvae, natural selection favors unspecific, general utilization of foliage as oviposition sites. Clumping of eggs then accelerates the oviposition process and has little adverse effect on survival of progeny because larvae forage for themselves; eggs are cryptic and available to carnivores when insects and other arthropods are least active. Proovigenic development of eggs is then adaptive, and rapid deposition of eggs fully formed from stored nutrients in larvae and pupae makes adult feeding superfluous. Larvae also evolve with a general capacity to feed broadly to some extent even when very small – in buds, in male cones, or within needles. As they grow a broad use of all foliage on a tree develops under crowded conditions. With such general feeding plus passive dispersal, wide host diet breadth is likely to evolve to include locally abundant conifer species. Because larval dispersal is so passive, a catholic diet is certainly adaptive in mixed conifer stands that prevailed in eastern Canada and the northeastern United States. Larvae had a greater probability of landing on a food plant even if it was not the optimal host species.

The resultant emergent properties are extensive (Fig. 7.2). A rather general diet including several conifer species means that resources in a northern forest are abundant and usually not limiting. The carrying capacity of the environment is very high, especially for small organisms of less than 100 mg wet weight at the fully fed, pupal stage. Then, populations can increase to very high densities under favorable conditions (1 in Fig. 7.2). Also, competition among females for oviposition sites is absent, and needles for egg laying are almost unlimited (2 in Fig. 7.2). In western spruce budworm females only avoid ovipositing with conspecifics at high densities (Leyva *et al.* in press). Crowding in females does not promote dispersal except under conditions of heavy defoliation and very high densities. With many females remaining in their local habitat, populations can build up to high densities. Abundant food, in the form of budworm larvae, attracts many predators, showing functional and numerical responses to larval densities (e.g. Mook 1963). For example, evening grosbeak populations reached staggering numbers during budworm outbreaks in New Brunswick, and breeding was very successful (Blais and Parks 1964; Blais and Price 1965). Also, larval abundance is high enough for epidemic diseases to become significant mortality factors. These were typically understudied and underestimated in budworm research because sampling was not frequent enough to catch brief and episodic pathogenic mortality. Another emergent property is the lack of larval competition over a very large range of densities when the carrying capacity per tree and in a forest is very high (3 in Fig. 7.2). In unsprayed areas of New Brunswick during epidemic budworm conditions samples of small larvae ranged from 12 to 2472 per 10 sq ft of foliage! (Miller 1958; Mott 1963). Populations can build to high densities, and only then will trees be defoliated and killed. Finally, when oviposition sites are abundant and females tolerate considerable crowding, for they oviposit up to 50 percent of their eggs before flying, female foraging is reduced, which contributes to the rapid increase in local populations each generation (4 in Fig. 7.2). These four lines of influence converge to result in eruptive population dynamics, almost inevitable at one time or another. At least the evolutionary background and the ecological conditions indicate an expectation for eruptive population dynamics.

The contrast between the hypothesis on spruce budworm and the gall-inducing sawfly, *Euura lasiolepis*, is striking. Comparing Fig. 7.2 for budworm with Fig. 3.10 for the sawfly, and adding the emergent properties for the sawfly, combined in Fig. 7.3, we see the full range of consequences based on divergent phylogenetic constraints. In Fig. 3.10 the evolution of high specificity in female behavior is illustrated and in Fig. 7.3 natural selection for high-quality site use is the key character that dictates the emergent properties discussed in Chapter 4, resulting in latent population dynamics. When females are highly specific in their oviposition preference, utilizing only modules of top quality, often coupled with tight linkage to larval performance, the distribution, abundance, and dynamics are dictated by a shortage of adequate modules. Patchy distributions, low abundance on a landscape, with pockets of relatively high density perhaps, and latent population dynamics are the consequences. The other extreme, illustrated by the spruce budworm, is when females are rather general in their oviposition decisions, tolerating a wide range of site types and qualities,

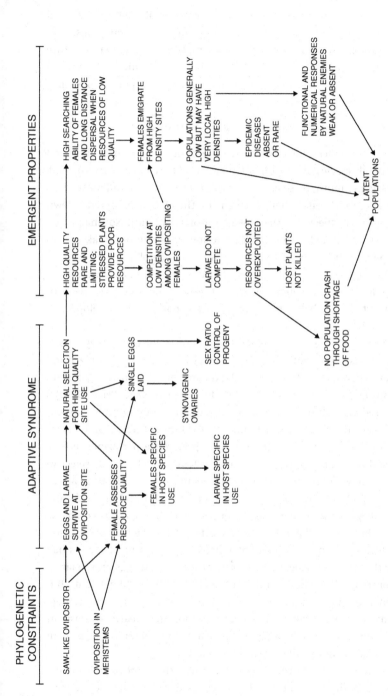

Fig. 7.3. The flow of influences in the stem-galling sawfly, *Euura lasiolepis*, for comparison with Fig. 7.1, resulting in latent population dynamics. (From Price 1990.)

there is almost no linkage to the quality of larval feeding sites, and larvae forage freely and independently from parental decisions. Outbreaks are likely to result and, with simple conifer forest structure widespread over a landscape and regionally, epidemics are likely to result in extensive tree death, economic impact, and public concern.

THE GENERAL CASE IN FOREST LEPIDOPTERA

Here in the spruce budworm we see the basic traits commonly found in outbreak species of forest lepidopterans that are defoliators: the budworms, gypsy moth, tussock moths, winter moth, brown-tail moth, mottled umber moth, and others. Focusing on the Macrolepidoptera known to be outbreak species in the forests of North America, Nothnagle and Schultz (1987) noted 41 species. Their criterion for an outbreak species was one that had caused visually conspicuous defoliation for at least two years in the 20 years of surveys from 1962 to 1981 in the U.S.A. These species form an independently and objectively identified sample with which to test the phylogenetic constraints hypothesis, as developed for the spruce budworm.

For all 41 species I searched the literature for life history details needed to test the hypothesis: place of oviposition, clutch size, place and time at which larvae started to feed, and any evidence for or against an ovipositional preference and larval performance linkage (Price 1994a). I discovered enough information on 33 species, representing 80 percent of the eruptive macrolepidopterans in U.S. forests (Table 7.1). In every species the evidence suggested lack of a preference–performance linkage and lack of an adaptive advantage for females to be highly selective during oviposition. Many species laid eggs in masses, indicating a lack of detailed choice on a per-egg basis, and many were relatively fecund with 100 eggs or more per cluster. Females with limited or no flight capacity represented 27 percent of the species (cf. Fig. 2.8 on the evolution of flightlessness in forest lepidopterans) and 33 percent of species had larvae that fed gregariously. A majority of species laid eggs off the larval food on twigs, branches, trunks, cocoons, or in litter (58 percent). But the most commonly noted trait, with 73 percent of species represented, was that eggs were laid in clusters. This indicates clearly a prevalent lack of any selective advantage to the careful placement of eggs as in *Euura lasiolepis* and similar kinds of species.

These results are certainly consistent with the phylogenetic constraints hypothesis for outbreak species, as developed in the scenario for the spruce budworm (Fig. 7.2). The consistency of the results is

Table 7.1. *Characteristics of Macrolepidoptera known to have shown outbreak population dynamics in the United States and reasons for lack of a tight preference–performance link between ovipositing females and larvae*

Family/species[a]	Common name	Oviposition	Lack of preference-performance link
Nymphalidae			
1. *Nymphalis californica*	California tortoiseshell	Adults overwinter and oviposit in spring	Oviposition on twigs.
Pieridae			
2. *Neophasia menapia*	Pine butterfly	5–20 eggs in a row	Eggs overwinter and hatch following spring.
Saturniidae			
3. *Coloradia pandora*	Pandora moth	5–50 eggs/cluster on needles or bark	Oviposition not specific to foliage, two-year life cycle. Gregarious feeders.
4. *Hemileuca nevadensis*	Nevada buck moth	Oviposition on twigs in clusters	Gregarious larvae.
Lasiocampidae			
5. *Malacosoma americanum*	Eastern tent caterpillar	200 eggs/mass on twigs	Overwinter as eggs, larvae feed on young foliage. Gregarious feeders.
6. *Malacosoma constrictum*	Pacific tent caterpillar	250 eggs/mass on twigs	Overwinter as eggs, larvae feed on new foliage. Gregarious feeders.
7. *Malacosoma disstria*	Forest tent caterpillar	170 eggs/mass on twigs	Overwinter as eggs, larvae feed on new foliage. Gregarious feeders.
8. *Malacosoma californicum*	Western tent caterpillar	170 eggs/mass on twigs	Overwinter as eggs, larvae feed on new foliage. Gregarious feeders.
9. *Malacosoma incurvum*	Southwestern tent caterpillar	Eggs in masses on twigs	Larvae feed on young foliage gregariously.

Table 7.1. (*cont.*)

Family/species[a]	Common name	Oviposition	Lack of preference–performance link
Lymantriidae			
10. *Orgyia pseudotsugata*	Douglas fir tussock moth	150 eggs/mass on female's cocoon	Females need not fly, larvae disperse by wind, overwintering as eggs, larvae feed on new foliage.
11. *Lymantria dispar*	Gypsy moth	750 eggs/mass on trunks and limbs	Females do not fly, larvae disperse by wind, overwinter as eggs, larvae feed on new foliage.
12. *Leucoma salicis* (= *Stilpnotia*)	Satin moth	650 eggs/mass on trees and other objects	Young larvae feed on old foliage, spin hibernaculum, then feed on new foliage.
13. *Dasychira plagiata*			Females with limited movement because of heavy bodies.
14. *Dasychira grisefacter*	Pine tussock moth	300 eggs/mass in loose clusters on needles	Larvae feed, hibernate under bark scales and then feed in spring. Females heavy-bodied with limited movement.
Notodontidae			
15. *Nygmia phaeorrhoea*	Brown-tail moth	Eggs in mass on under-side of leaf	Females heavy-bodied. Young larvae gregarious, feed in fall and spring.
16. *Datana integerrima*	Walnut caterpillar	120 eggs/mass on under-sides of leaves	Larvae molt on trunk of tree so female does not select most of foliage in larval diet.

Table 7.1. (*cont.*)

Family/species[a]	Common name	Oviposition	Lack of preference–performance link
17. *Heterocampa guttivitta*	Saddled prominent	Up to 500 eggs singly on leaves	Larvae migrate frequently from tree to tree.
18. *Heterocampa manteo*	Variable oak leaf caterpillar	Up to 500 eggs laid singly on leaves	Female does not have time to be very selective?
19. *Symmerista canicosta*	Red-humped oakworm	66 eggs/mass on leaves	Larvae are gregarious so feed well away from oviposition site.
Dioptidae			
20. *Phryganidia californica*	California oakworm	60 eggs/mass on leaves and elsewhere	Adults weak fliers and oviposition is nonspecific.
Arctiidae			
21. *Halisidota argentata*	Silver-spotted tiger moth	100 eggs/mass on twigs and needles	Unspecific oviposition, colonial feeders.
22. *Halisidota ingens*		Eggs on twigs	Larvae gregarious.
23. *Hyphantria cunea*	Fall webworm	600 eggs/mass on under-sides of leaves	Gregarious until last instar.
Geometridae			
24. *Anacamptodes clivinaria*	Mountain mahogany looper	350 eggs/mass in bark crevices	Oviposition away from feeding site.
25. *Alsophila pometaria*	Fall cankerworm	112 eggs/mass on smaller twigs and branches	Female wingless, oviposition away from feeding site.
26. *Ennomos subsignarius*	Elm spanworm	70 eggs/mass on twigs	Oviposition away from foliage.
27. *Lambdina fiscellaria*	Hemlock looper	Eggs single or in small groups on moss, lichens, bark on limbs and trunks	Overwinter as egg, feed on opening shoots.

Table 7.1. (*cont.*)

Family/species[a]	Common name	Oviposition	Lack of preference–performance link
28. *Lambdina punctata*		Eggs single in leaf litter and bark scales	Larvae feed on developing buds and leaves.
29. *Nepytia freemani*		Eggs in small clusters in late summer on needles	Eggs hatch in spring and larvae feed on new foliage.
30. *Nepytia phantasmaria*	Phantom hemlock looper	Eggs laid singly in fall	Eggs overwinter, larvae feed on new foliage.
31. *Operophtera bruceata*	Bruce spanworm	Eggs single in bark crevices, etc.	Female almost wingless, eggs over-winter and hatch in early spring.
32. *Paleacrita vernata*	Spring cankerworm	100+ eggs/mass laid in early spring under bark or in crevices, before foliage is out	Female wingless.
33. *Phaeoura mexicanaria*		160 eggs/mass on needles	Larvae feed indiscriminately on new and old foliage.

[a] Species are listed according to Nothnagle and Schultz (1987) and biological attributes are from W. L. Baker (1972) and/or Furniss and Carolin (1977).
Source: Price (1994b).

remarkable in my opinion, and compelling evidence that evolved characters of behavior and life history are critical for a predictive view of population ecology.

Among microlepidopteran moths many species show no ovipositional preference in relation to larval feeding habits, as we saw for the spruce budworm. Many are serious pests in forests, although shoot borers and cone worms were probably rare in natural forests and only became serious pests in managed monocultures. A very common life cycle is for eggs to be laid in the summer, and larvae hatch in late summer and may or may not feed before overwintering in a protected site, often a hibernaculum, after which they complete feeding in the spring. This pattern fits the *Choristoneura* species, *Archips* species, and *Acleris*, all in the Tortricidae. Pyralids in the genus *Dioryctria* show the same pattern, as do some olethreutids (*Zeiraphera*, *Gypsonoma*, and *Epinotia*), gelechiids (*Recurvaria*, *Exoteleia*, *Dichomeris*), coleophorids (*Coleophora*), and psychids (*Thyridopteryx*). In fact, in eastern North America 28 species with known life cycles fit this pattern where larvae feed in the spring while eggs are laid in late summer (cf. Baker 1972). Therefore, the phylogenetic constraints, adaptive syndromes, and emergent properties are likely to be equivalent to those of the spruce budworm. Many are outbreak species, although those that are still limited to large modules, such as the shoot and cone borers (e.g. *Dioryctria*), will be exposed to a low carrying capacity in natural forests.

The trait of overwintering as larvae is also common in European microlepidopterans. Of 34 common species in northern Finland, 30 have known overwintering stages, of which 93 percent overwinter as larvae (Kozlov *et al.* in press), probably involving spring feeding well detached in time and space from a parental female's decisions regarding egg laying. Therefore, we should not be surprised to observe that many species that feed on leaves, such as budworms, bud moths, webworms, leaf rollers, casebearers, and bagworms become, at times, serious pests in forests and woodlands.

THE GENERAL CASE IN BUTTERFLIES

Very few butterflies are outbreak species in forests. Two species are noted in Table 7.1, the California tortoiseshell and the pine butterfly. And very few species are outbreak species in other natural vegetation, although in agricultural monocultures a few species can become serious pests. Female butterflies are generally highly selective about where they lay eggs; eggs are laid on foliage on which larvae will commence feeding in a few days. There is a tight connection between oviposition and larval feeding site. And many species of butterfly are rare, endangered, near to extinction, and even extinct (cf. Arnett 1993; Gaston *et al.* 1993; Samways 1994; Thomas *et al.* 1998). All these are traits expected to correlate with latent population dynamics.

Of about 470 species of butterflies recorded in North America, only two are mentioned as pests by Arnett (1993), the southern cabbageworm (*Pieris protodice*) and the imported cabbageworm (*Pieris rapae*).

In their treatise covering more than 650 species of insect that damage trees and shrubs, Johnson and Lyon (1976) mention only one butterfly and one skipper: the morningcloak, *Nymphalis antiopa*, and the silverspotted skipper, *Epargyreus clarus*. We could add a few more species perhaps, but we are in the range of about 1 percent of species thought to be pest species, a very low percentage!

There are many characteristics of butterflies that render them very sensitive to habitat variation, host plant quality, and the presence of other butterflies (cf. Steffan-Dewenter and Tscharntke 1997). The general nature of habitats for butterflies was expressed by Thomas *et al.* (1998) in their studies on the population dynamics of *Maculinea* species, beautiful large blue butterflies in the family Lycaenidae. The lycaenids constitute the largest family of butterflies in North Temperate latitudes (136 species in North America alone, or 29 percent of butterfly species), with larvae living mutualistically with ants. Thus, in addition to a suitable food plant, a suitable ant species must be present nearby. The details of the interactions are illustrated in Fig. 7.4, and habitat quality variation is basic to understanding these butterflies. "*Maculinea* (large blue) butterflies are rare and specialized insects" (Thomas *et al.* 1998, p. 262).

> When measuring habitat, we began with Singer's (1972) premise that, narrow though the niches of many butterflies may be (Thomas, 1991), sites in real landscapes that support the larval food do not exist – as many metapopulation models assume – as homogeneous patches of universally suitable (source) or unsuitable (sink) habitat, which change over time only in their number, areas, distribution and the permeability to dispersal of intervening land. Instead, sites may contain a variety of types ranging, among sources, from habitat of optimum quality where r and K are the maximum possible for the species under a given climate, through a continuum of sub-optimal but suitable habitats producing fewer butterflies. Similarly, sink habitat varies from absolute sinks where mortality is total, to sub-areas where the butterfly's intrinsic rate of increase, though <1, contributes some adults to the population, and perhaps increases its probability of persisting through extreme years when normal source types of habitat become unsuitable (Sutcliffe *et al.* 1997). (Thomas *et al.* 1998, pp. 263–264)

This is precisely the habitat situation and population structure of gall-inducing sawflies in the genus *Euura* (Chapters 3 and 4). There may also be rapid evolution in butterfly populations in response to host plant availability, with independent shifts in different populations

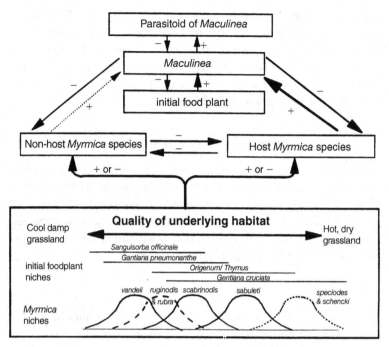

Fig. 7.4. The kinds of organisms and habitats that interact with five *Maculinea* butterfly species, defining narrow niche occupation and patchy distribution. Each species of butterfly must find a location with its initial food plant present, plus a specific *Myrmica* ant species which will act as a host for larvae. In addition host and nonhost ant species interact negatively and parasitoids may enter ant nests or attack larvae on food plants. (Reproduced from Thomas, J. A., R. T. Clarke, G. W. Elmes, and M. E. Hochberg (1998) Population dynamics in the genus *Maculinea* (Lepidoptera: Lycaenidae), Fig. 11.1, in J. P. Dempster and I. F. G. McLean (eds.) *Insect populations: In theory and in practice*, with kind permission of Kluwer Academic Publishers.)

(Singer *et al.* 1992, 1993; Radtkey and Singer 1995; Singer and Thomas 1996).

As in *Euura* sawflies, butterflies show well-correlated linkage between oviposition preference of adult females and larval performance in some cases but by no means all (cf. Singer 1971; Rausher 1982; Singer *et al.* 1988; Courtney and Kibota 1990; Thompson and Pellmyr 1991). And many butterfly species are territorial (e.g. R. R. Baker 1972, 1983; see also Shapiro 1970), (cf. Figs. 2.9 and 2.10), females avoid ovipositing near the eggs of conspecifics (and mimics of eggs; Gilbert 1975, 1991), and space eggs individually and extensively (Cromartie 1975a, b). Young plant parts, meristems and new tendrils, may be used as oviposition sites by some butterfly species, and females can be seen to spend many minutes searching plants for acceptable sites. Many species also use herbaceous plants as larval food and adult nectar sources, species typically very patchy over a natural landscape.

PEST STATUS IN OTHER ARTHROPOD GROUPS

Many other pairwise comparisons, as with the moths and butterflies, can be made between related taxa, one showing outbreak characteristics or pest status and the other showing more latent population dynamics with only rarely found pest species among them. These, and the reasons for the differences, will be treated as we work down the orders of arthropods with phytophagous groups included.

Mites

Mites, in the Order Acari, include the spider mites, Family Tetranychidae, which are phytophagous and are frequently serious pests. Mites spin a silken thread as they travel over plants, and when crowded they move up a plant to the tip and disperse passively on the silk. Under such conditions, when landing on plants is uncontrolled, immatures are likely to evolve with a general capacity to feed on many plant species, as seen in the two-spotted spider mite, *Tetranychus urticae*, a notorious pest in glass houses, on house plants, and outdoors. "There are very few plants grown in greenhouses which are not subject to injury by these spider mites" (Metcalf and Metcalf 1993, p. 18.20). Where dispersal is passive, and mites are too small to select among plants, little or no preference evolves; generalized feeding habits evolve with the consequence of an evolutionary potential for becoming a serious pest species with eruptive population dynamics. A scenario similar to the spruce budworm and species listed in Table 7.1 is evident.

Stick insects

Phasmids, or stick insects, include several species that are serious defoliators of *Eucalyptus* forests in Australia. Female adults are flightless and drop eggs from high in eucalyptus trees; eggs fall to the ground and some are collected by ants and carried into nests. After 6–18 months on the forest floor, eggs hatch in the spring and nymphs search for a host tree to climb. "Searching is a random process of trial and error, the nymph climbing any projecting object (e.g. plant stem) that it happens to encounter and settling eventually on the first sapling or large tree bearing food" (Readshaw 1965, p. 477). Nine species of *Eucalyptus* are fed on extensively by one of the species *Didymuria violescens*. The other two plague species in Australia are *Podocanthus wilkinsoni* and *Ctenomorphodes tessulatus*. The existence of kentromorphism, in which phase differences occur, in all three species suggests that they have a long evolutionary history of outbreak dynamics (Key 1991). At low densities nymphs are green and cryptic, while at high densities nymphs become conspicuous and aposematic with black, yellow, and sometimes white patterning. The phasmids thus appear to resemble the flightless forest Lepidoptera in their phylogenetic constraints, adaptive syndromes, and emergent properties (cf. Fig. 2.8). But the rift between female oviposition and larval feeding is extreme in the stick insects: eggs are dropped haphazardly to the ground and nymphs search for food at random. In addition, eggs lie dormant from 6 months to 1.5 years on the ground and may be carried by ants. Clearly there can be no ovipositional preference in relation to larval food quality and larval performance; larvae will evolve to be rather general feeders, at least on *Eucalyptus* species, and eucalypt forests provide a very high carrying capacity for nymphs. All contribute to the probability of eruptive population dynamics.

Orthoptera

Among the Orthoptera there is the interesting comparison to be made between the extremely eruptive short-horned grasshoppers (Acrididae) and the bush katydids in the long-horned grasshoppers (Tettigoniidae: Phaneropterinae). Short-horned grasshoppers include the plague locusts, highly eruptive species at least since biblical times, and many other damaging grasshopper species. "Few, if any, other species of insects have caused greater direct loss to crops than have grasshoppers" (Metcalf and Metcalf 1993, p. 94, referring to acridids). Of the 54 species of grasshoppers listed as pests outside North America and temperate Eurasia, 80 percent belonged to the family Acrididae (Anonymous 1982). All species were regarded as sufficiently injurious to crops that control measures were justified. Among the bush katydids, the most diverse subfamily of tettigoniids in Australia with several hundred species present, very few are considered to be pests (cf. Rentz 1996). I will distinguish these groups as grasshoppers for the Acrididae and katydids for the tettigoniid subfamily Phaneropterinae. Both groups are predominantly plant feeders, so we should wonder why the grasshoppers are

Fig. 7.5. The life cycle of a typical acridid grasshopper including locust species. Eggs overwinter in the ground in a pod or ootheca. Nymphs hatch in the spring and wiggle to the surface, pass through six instars, and molt into winged adults. Females oviposit into the ground, boring deeply in some species, with short sturdy "blades" that form the ovipositor. (From Pfadt 1988.)

such notorious pests and are well studied, while the katydids are poorly studied because of their nonpest status. "It must be emphasized that little or nothing is known of the biology and behavior of the majority of species" (Rentz 1996, p. 110). This, of course, makes comparisons difficult, but I think a clear-cut case can be made for fundamentally different phylogenetic constraints in the two groups.

The life cycle of a typical grasshopper starts with eggs in the soil, special adaptations for nymphs emerging from the soil in spring (Bernays 1971a, b, 1972a, b, c), nymphs feeding generally on grasses and forbs, with winged adults mating and females ovipositing clutches of eggs in the soil (Fig. 7.5). One generation per year is usual in temperate regions but three or more generations may occur in locusts such as the desert locust, *Schistocerca gregaria*.

Female grasshoppers tend to oviposit in open ground away from vegetation, depositing eggs in clutches or pods often covered by a hardened foamy coat or ootheca. Clutch sizes may reach 60–85 eggs in some species. Thus, nymphs emerging from their eggs must ease their way

Table 7.2. *Highly eruptive grasshopper species from around the world*

Species	Common name	Location
Anacridium melanorhodon	Tree locust	Africa
Austriocetes cruciata	Plague grasshopper	Australia
Chortoicetes terminifera	Australian plague locust	Australia
Dociostaurus maroccanus	Moroccan locust	Western Turkey
Gastrimargus musicus	Yellow-winged locust	Australia
Locusta migratoria	African migratory locust	Africa, Asia, Australia
Locustana pardelina	Brown locust	Southern Africa
Melanoplus sanguinipes	Migratory grasshopper	North America
Melanoplus spretus	Rocky Mountain grasshopper	North America
Nomadacris guttulosa	Spur-throated locust	Australia
Nomadacris septemfasciata	Red locust	Tropical Africa
Nomadacris succincta	Bombay locust	India
Schistocerca gregaria	Desert locust	North Africa, Asia

Source: From Price, P. W. (1997) *Insect ecology*, 3rd edn, Table 19.2, © 1997, reprinted by permission of John Wiley & Sons, Inc.

to the soil surface, travel a distance to the nearest foliage, and start to feed. Oviposition behavior is separated significantly in time and space from larval food, with the now familiar expectation that there will be no evolved ovipositional preference in relation to nymphal food, and progeny are liable to be general in their abilities to feed on a wide variety of food plants (cf. Chapman and Sword 1997).

Densities of eggs in soil can reach extremes of 780 egg pods per sq m or 103 740 eggs per sq m in the desert locust, with soil having a very high capacity as an oviposition medium. Unusually heavy rains in desert areas increase the carrying capacity of plants for nymphal and adult feeding and populations can increase rapidly over several generations, densities increase dramatically, a phase change from solitary to gregarious occurs, and a plague is born (Showler 1995).

The plague locust syndrome has evolved many times with numerous genera involved around the world (Table 7.2), added to which are dozens of eruptive grasshoppers without phase change that defoliate natural vegetation and agricultural crops. Hence, efforts to control grasshopper populations have been long term and are ongoing (e.g. Govindachari and Suresh 1997; Lockwood and Ewen 1997; Riegert *et al.* 1997). Worth noting is the evidence suggesting that the Rocky Mountain "locust," *Melanoplus spretus*, is now extinct (Lockwood 2001).

Such boom-and-bust dynamics apparently render species vulnerable to relatively limited habitat alteration.

The key features of the grasshopper phylogenetic constraints, adaptive syndrome, and emergent properties are summarized in Table 7.3, and the flow of effects from evolved characters to the potential for eruptive population dynamics is illustrated in Fig. 7.6.

The bush katydids (Phaneropterinae) are about as different from the short-horned grasshoppers as it is possible to imagine for two kinds of phytophagous insects in the same order, Orthoptera (Table 7.3, Fig. 7.7). "The ovipositor of many species is short, broad and laterally compressed and armed. Its function is to split the edge of leaves, into which the hard, black, disk-like eggs are laid. A large number utilize this form of oviposition" (Rentz 1996, p. 110). Hence, the phylogenetic constraints of the sawflies, discussed in Chapters 3 and 5, and bush katydids are convergent. Another convergent group is the membracids discussed in Chapter 6 (see also Price and Carr 2000). For the fauna of southern Africa, Scholtz and Holm (1985) note that the bush katydids include many endemic genera and species "are usually closely associated with specific host plants or plant communities . . . Almost all species are nocturnal and lay their eggs in plant tissue, usually by inserting the flat ovipositor into the edges of leaves, or in stems or twigs" (p. 83).

Rentz (1996) noted that the bush katydids are the most diverse group in the Australian tettigoniid fauna, with several hundred species represented, but wondered why this group had speciated so much more than other groups of tettigoniids. There are relatively few genera but many species per genus (Rentz 1991), suggesting rapid speciation. My prediction is that the ovipositor and associated behaviors demand considerable specificity in host plant species utilization with a probability of host plant shifts in sympatry and consequent speciation (cf. Bush 1975a, b; Bush and Smith 1998).

The Phylogenetic Constraints Hypothesis appears to be applicable to the grasshoppers and katydids. Very different phylogenetic constraints result ultimately in divergent emergent properties.

Homoptera

Aphids and scale insects are very serious pests and show eruptive population dynamics on trees, shrubs, and herbaceous plants (cf. Johnson and Lyon 1976; Metcalf and Metcalf 1993). Hundreds of species in each group are serious pests, but the extent to which they would be highly eruptive in natural landscapes is open to debate.

Table 7.3. *Comparison between short-horned grasshoppers (Acrididae) and bush katydids (Tettigoniidae: Phaneropterinae)*

Character	Acridids	Katydids
1. Order	Orthoptera	Orthoptera
2. Suborder	Caelifera: short-horned	Ensifera: long-horned
3. Family	Acrididae	Tettigoniidae
Phylogenetic constraint		
4. Oviposition and ovipositor structure	Into soil, with short robust ovipositor and greatly extendable abdomen	Into plant parts with short broad and laterally compressed ovipositor
Adaptive syndrome		
5. Oviposition site	Warm mineral soil in open ground	Plant modules in woody vegetation
6. Egg protection	Egg in ootheca in soil	Eggs in plant tissue
7. Clutch size	Frequently large, 60–85 eggs	Single eggs or small groups
8. Nymphal characters	Specialized traits for emerging from soil	?
9. Male parental investment	Moderate	High
10. Preference–performance linkage	Absent	Predicted to be high
Emergent properties		
11. Resource supply – substrate for eggs	Can be very high/limitless	Probably low
12. Resource supply – food	High carrying capacity	Low carrying capacity
13. Population through time	Variable – depending on variable weather	Predicted to be stable
14. Population maxima	Very high (e.g. 103 740 eggs/m^2)	Low
15. Population dynamics	Eruptive	Latent
16. Frequency as economically important	High	Low

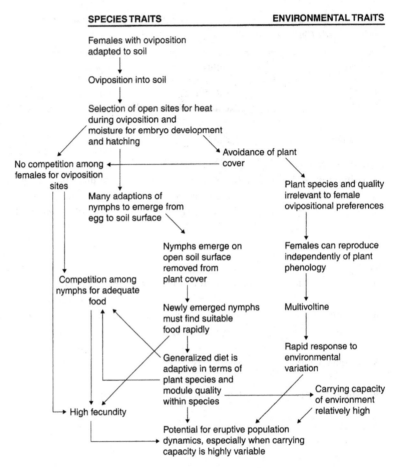

Fig. 7.6. The flow of influences in acridid grasshoppers from the phylogenetic constraint of oviposition into the soil to the potential for eruptive population dynamics.

Aphids are not speciose relative to many groups of herbivorous insects, with low species diversity explained by Dixon *et al.* (1987) because "of the constraints imposed by their way of life, namely, the short period for which they can survive without food, their high degree of host specificity, and the low efficiency with which they locate host plants" (p. 590). "Parthenogenesis and feeding on phloem sap both developed early in the evolution of aphids. Parthenogenesis led to the telescoping of generations, such that embryos contain the next generation of embryos. This simultaneous commitment to growth and reproduction puts a severe constraint on the length of

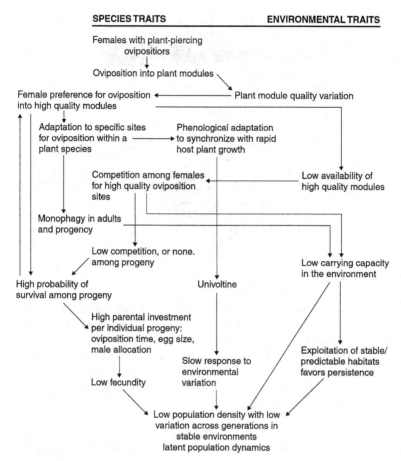

SPECIES TRAITS   **ENVIRONMENTAL TRAITS**

Fig. 7.7. The phylogenetic constraint of bush katydids, with plant-piercing ovipositors, to the emergent property of latent population dynamics.

time aphids can survive without feeding and, thus, the time they can spend searching for host plants. Small size and host specificity, both adaptations to feeding on phloem sap, also impose constraints" (Dixon 1994, p. 581) (Fig. 7.8). This statement by Dixon *et al.* is exactly the kind of approach adopted in the Phylogenetic Constraints Hypothesis. Parthenogenesis and abbreviation of generations, phylogenetically basic to the aphids with an origin perhaps 200 million years ago, act as phylogenetic constraints. Of course they compensate with an adaptive syndrome including selection of high-quality feeding sites as far as possible, both rapidly growing shoots and leaves and senescing modules where transport of nutrients out of tissues is rapid. Hence,

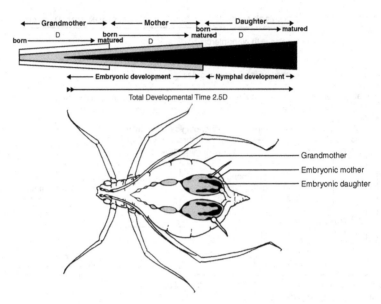

Fig. 7.8. Dixon's (1994) explanation of the telescoping of generations of aphids with embryonic development of granddaughters starting in the grandmother. D indicates the time from birth to the beginning of reproduction, about 7 days under favorable conditions. This is very rapid for an organism of its size. However, an individual has already developed as an embryo for 1.5 D in the mother, making total development time 2.5 D or about 17–18 days.

they fit White's terms "flush and senescence feeders" in general although some species may be specialized on flushing modules, others on senescing modules, and still others on both (cf. Dixon 1973, 1994; White 1993). The combination of the phylogenetic constraints and the adaptive syndrome results in spectacular clonal growth on high-quality modules with the potential for serious damage to host plants. The phylogenetic constraints on aphids are unique to the group, requiring a special case to be made in relation to the Phylogenetic Constraints Hypothesis.

Scale insects and mealybugs (Homoptera: Coccoidea) conform more closely to the general situation for outbreak species with a very passive, often long-distance, dispersal of first instar, wingless crawlers. There cannot be a female ovipositional preference in relation to nymphal food because she is wingless and immobile. The life history becomes the phylogenetic constraint for the group, with sessile adult females and highly mobile crawlers. No preference–performance linkage is possible and the haphazard movement of crawlers has, no doubt, resulted in selection for broad host plant use, or extensive adaptive radiation of specialists. More than 6000 species are described worldwide and double this number probably exist (Williams 1991). Some species reach such high populations that large trees can be killed, and this threat is widespread on many host plant species, imposed by coccids such as the oystershell scale, *Lepidosaphes ulmi*. This species is an amazing generalist with perhaps 150 known hosts, and locally may infest plants in 12 families and 19 genera (Johnson and Lyon 1976).

A pairwise comparison is instructive among homopteran families, Membracidae, or tree hoppers, Aphrophoridae and Cercopidae. The last two families are called froghoppers or spittlebugs. They are closely related families but some divergent patterns exist in certain cases. The membracids were mentioned in Chapter 6 as an excellent example of convergence with the *Euura lasiolepis* model. Four species were listed in Table 6.2. And the froghopper *Aphrophora pectoralis*, fondly known as the "awafuki" in Japan, also shows correlated female ovipositional preference and larval performance (Craig and Ohgushi in press). Awafuki females oviposit into stems, as in many membracids and sawflies, but in this case the stems are killed by the many scars, so nymphs move to new shoots in the spring. Nevertheless, females select vigorous shoots on which to feed and oviposit, and nearby shoots to which nymphs move in the spring are equally vigorous, or even more so as progeny are mobile and show their own preferences. Awafuki can become locally abundant but favorable sites are very patchy according to a survey in Hokkaido by Craig and associates (T. P. Craig et al., unpublished data).

Why then, are several genera of spittlebugs such serious pests of forage grasses in tropical latitudes? The immediately apparent difference from *Aphrophora* in Japan is that species oviposit in the soil, evidently an old trait in the grass-feeding spittlebugs (Pires et al. 2000). Several genera are involved: *Aeneolamia*, *Deois*, *Mahanarva*, *Prosapia*, and *Zulia*. All species are known to oviposit into the soil, while some will utilize plant debris or dead leaf sheaths, but no living plant tissue is used. Hence, small nymphs, newly hatched from eggs in the soil after the prolonged winter drought, just forage for themselves. They are unable to travel far, without time to exercise choice among plant species or plant parts. In detailed studies on *Deois flavopicta*, near Brasilia, females did prefer to oviposit near plants and in moist soils, but they did not discriminate among plant species or plant quality (Pires 1998; Pires et al. 2000). Here is another case, then, where a phylogenetic constraint in the life history, oviposition in soil, causes a separation of female oviposition behavior and nymphal food quality. Generalized feeding capacity in nymphs is then adaptive, resulting in a broad range of food items, a high carrying capacity, and the potential for eruptive population dynamics. All reminiscent of the acridid grasshoppers.

Coleoptera

Many beetle species in the family Chrysomelidae, the leaf beetles, share the same traits as tropical cercopids: they lay eggs in the soil and small larvae must forage for themselves. Two subfamilies, the flea beetles, Alticinae, and the root worms and others in the Galerucinae, include many serious pest species in agriculture and on woody plants. It is rare for leaf beetles to kill woody plants, but flea beetles can. Our research group studied a large flea beetle, *Disonycha pluriligata*, which we found defoliating and killing coyote willow, *Salix exigua*, near Flagstaff, Arizona, in a more-or-less natural setting (Dodge *et al.* 1990; Dodge and Price 1991a, b; Marques *et al.* 1994). The same causal pathway in eruptive species, as we have seen in this chapter, is evident here: small larvae forage for themselves, even though specific to the host plant species; they feed on the first foliage they can find and are generally capable of feeding on all foliage as they become mature. Defoliation results and death of clones can occur. In fact, in the population we studied near Flagstaff, both the beetle and the willow are now extinct, as the result of overexploitation of resources by the beetle. Flea beetles are often polyphagous, being serious pests on corn, millet, sorghum, sweet potatoes, the cabbage family, and the potato family. Among the Galerucinae are included "some of the world's most destructive insects" (Metcalf and Metcalf 1992). Sixty-four species from many parts of the globe are listed as pests by Metcalf and Metcalf (1992), all associated with the cucumber family (Cucurbitaceae) in one way or another, but larvae feed on roots of very important crop plants: cucurbits, sweet potato, corn, and soybeans. Females may lay 500–1000 eggs individually in the ground at the bases of plants, and larvae tunnel into roots and kill them in the process, causing the plants to topple. Oviposition in the soil leaves females unable to evaluate host plant quality, with the now familiar results.

Bark beetles (Scolytidae) are a major scourge in coniferous forests around the world, but very few species have broken out of the phylogenetic mold of utilizing senescent, moribund trees or tree parts. In North America about 98 percent of the scolytid species (470 out of 480 species) breed in the cambial tissues of dying trees or tree parts or in cones. But a few have escaped from these habitats and attack and

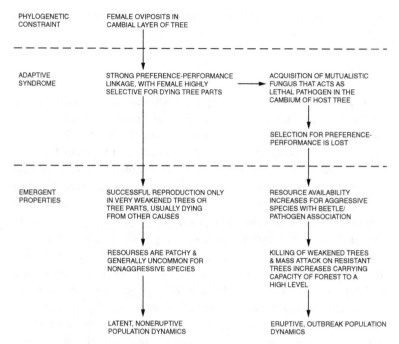

PHYLOGENETIC   FEMALE OVIPOSITS IN
CONSTRAINT   CAMBIAL LAYER OF TREE

ADAPTIVE   STRONG PREFERENCE-PERFORMANCE   ACQUISITION OF MUTUALISTIC
SYNDROME   LINKAGE, WITH FEMALE HIGHLY   ⟶   FUNGUS THAT ACTS AS
                     SELECTIVE FOR DYING TREE PARTS   LETHAL PATHOGEN IN THE
                                                        CAMBIUM OF HOST TREE

                                                        SELECTION FOR PREFERENCE-
                                                        PERFORMANCE IS LOST

EMERGENT   SUCCESSFUL REPRODUCTION ONLY   RESOURCE AVAILABILITY
PROPERTIES   IN VERY WEAKENED TREES OR   INCREASES FOR AGGRESSIVE
                     TREE PARTS, USUALLY DYING   SPECIES WITH BEETLE/
                     FROM OTHER CAUSES   PATHOGEN ASSOCIATION

                     RESOURSES ARE PATCHY &   KILLING OF WEAKENED TREES
                     GENERALLY UNCOMMON FOR   & MASS ATTACK ON RESISTANT
                     NONAGGRESSIVE SPECIES   TREES INCREASES CARRYING
                                                        CAPACITY OF FOREST TO A
                                                        HIGH LEVEL

                     LATENT, NONERUPTIVE   ERUPTIVE, OUTBREAK POPULATION
                     POPULATION DYNAMICS   DYNAMICS

Fig. 7.9. The phylogenetic constraints hypothesis applied to noneruptive and eruptive species of bark beetles (Scolytidae). (From Price, P. W. (1997) *Insect ecology*, 3rd edn, Fig. 19.9, © 1997, reprinted by permission of John Wiley & Sons, Inc.)

kill healthy trees (e.g. Berryman 1982; Paine *et al.* 1997). The original condition for scolytids was that females showed a strong preference for dying or heavily weakened trees in which reproduction was successful. A well-developed preference–performance linkage prevailed. In natural forests available breeding sites would be very patchy, with isolated old trees becoming susceptible to attack, but during droughts or after fires carrying capacity would increase briefly. Populations would remain low and very patchy in most landscapes. The emergence of highly aggressive, tree-killing bark beetles appears to have resulted from the acquisition of mutualistic fungi pathogenic to the host trees plus mass attack on healthy trees (e.g. Berryman 1982; Flamm *et al.* 1988; Raffa 1988). The result is to increase the carrying capacity of a forest enormously, and to provide an evolutionary basis for potentially eruptive species of bark beetle in the genera *Dendroctonus* and *Ips* (Fig. 7.9). While this view over-simplifies a diverse set of complex interactions (e.g. Paine *et al.* 1997), there is no doubt about the unique role of these beetles killing trees and the intimate mutualistic relationship with fungal pathogens of the trees (cf. Whitney 1982).

Lepidoptera

In addition to the forest Lepidoptera discussed early in this chapter (cf. Table 7.1), many other species illustrate eruptive population dynamics. Just a few additional examples will be treated here, especially using tropical species, for they illustrate the same kinds of patterns we have seen in north temperate regions. Mopane woodland straddles the Tropic of Capricorn in Africa and provides food for large caterpillars called "mopane worms." They are larvae of a magnificent emperor moth, *Gonimbrasia belina* (Saturniidae). The caterpillars become extremely abundant, defoliating trees, but providing an important protein source for native Africans. The 5.5 cm caterpillars are squashed to remove gut contents, dried, and used in various nutritious recipes, which I have tried. I must say that the dried cadavers one can buy by the kilo in Johannesburg have not proved popular with my students, but rehydrated and properly prepared they are no doubt splendid (cf. Skaife *et al.* 1979; Scholtz and Holm 1985; Menzel and D'Aluisio 1998; Green 2001). Other emperor moths in southern Africa are also serious defoliators, including on commercially important trees, such as the pine tree emperor, *Imbrasia cytherea*, and the cabbage-tree emperor, *Bunaea alcinoe*.

The mopane worm adult female is unspecific in its oviposition, laying eggs on bark or leaves (Scholtz and Holm 1985) and larvae are generally capable of eating all foliage on a mopane tree. Hence, mopane woodland, which tends to be monospecific in many areas, provides a high carrying capacity for larvae, which in turn become a significant protein source for large numbers of people. Unfortunately, the population dynamics of this species have not been studied enough to understand the sporadic eruptions. Note that in Table 7.1 two North American saturniid moths with outbreak population dynamics share with the mopane worm similar life history traits.

Several tropical snout moths (Pyralidae) have shifted from native plants to grass crops, such as sugar cane and corn, becoming serious pests. Such shifts have been facilitated, no doubt, by the unspecific nature of female ovipositional behavior in relation to food quality for the progeny. For example, the paper plant or papyrus, *Cyperus papyrus*, is a natural host to a pyralid moth, now called *Eldana saccharina*, because of its serious pest status on sugar cane. Papyrus produces umbels with brown papery bracts at the bases of the mature umbels which form the substrata for oviposition by *Eldana*. Larvae hatch and feed in the base of the umbel and then feed in the cortex of the rhizome (Conlong 1990). *Eldana* was first reported on sugar cane in South Africa in 1945. It lays eggs on sugar cane on dead foliage, an equivalent site to dead bracts of papyrus, but sugar cane, as it grows, provides vastly more oviposition sites than papyrus. Other pyralids include the sugar cane borer, *Diatraea saccharalis*, and two borers on corn, *D. crambidoides* and *D. grandiosella*. Eggs are laid on leaves, with clutches ranging from 2 to 100 in *D. saccharilis*, with a total of up to 300 or 400 eggs noted for the corn borers (Metcalf and Metcalf 1993). Clearly, females are not being selective when so many eggs are laid, often in large clutches. And the lesser cornstalk borer, *Elasmopalpus lignosellus*, is actually fire-adapted in its native habitat, with adult females laying eggs onto the youngest new sprouts of grasses as they grow from a burned clump. In the cerrado vegetation of Brazil, with frequent fires, no doubt *Elasmopalpus* females made an easy shift from new shoots of native grasses to the new shoots of corn on which eggs are laid. The common stalk borer of corn, *Papaipema nebris*, is a noctuid that lays eggs in late summer and fall unspecifically in grasses and weeds, with females producing more than 2000 eggs in some cases. Larvae hatch in the spring and bore into grass stems, shifting from stem to stem as they grow and seeking out larger stems, like corn, in later instars. This life cycle shares many features with the spruce budworm example used early in this chapter. In fact noctuids, including the cutworms, armyworms, bollworms, loopers, and others, are serious pests in both temperate and tropical latitudes. In many cases larvae move extensively among plants or they hide off plants during the day under stones and debris. The eruptive dynamics of various species and their extensive migratory behavior have prompted comparisons with the plague locusts (e.g. Betts 1976; Rainey 1979; Drake and Gatehouse 1995). Armyworms also show a phase change from solitary to gregarious, prompting further comparisons with locusts (e.g. Rose 1975; Rose and Khasimuddin 1979; Dingle 1996). In the United States, the beet armyworm, *Spodoptera exigua*, has been recorded from more than 50 plant species in 18 families and the fall armyworm, *S. frugiperda*, has been found on more than 60 plant species (e.g. Mitchell 1979).

When we compare lists of defoliating insects from the north temperate and the tropics, the range of lepidopteran families is remarkably similar. For example, Wagner *et al.* (1991) listed the defoliating insects of Ghanaian forests, which include the following families also represented in Table 7.1: Nymphalidae, Saturniidae, Lasiocampidae, Lymantriidae, Notodontidae, Arctiidae, and Geometridae. Unfortunately, life histories of representative species in the tropics are not adequately studied for comparisons with traits noted in Table 7.1. Wagner *et al.* (1991) also listed pyralid moths, tortricid moths, and acridid grasshoppers as defoliators in Ghana, all discussed in this chapter.

Hymenoptera

An interesting contrast exists between the tenthredinid sawflies covered in Chapters 3–5, which are not regarded generally as eruptive or pest species, and the diprionid sawflies on conifers, many of which are serious defoliators. This comparison is developed in Chapter 8.

THE CONTINUUM FROM LATENT TO ERUPTIVE POPULATION DYNAMICS

I have emphasized in the preceding pages the extremes in a continuum from species with latent population dynamics to those that are eruptive, destructive pests. Clearly, many species exist on the continuum, between the extremes, and why they do is more difficult to discern. They provide a major challenge in any approach to population dynamics, and especially for an evolutionary hypothesis. There are also many species that we might expect to be at one extreme or the other, but they do not fit the predictions of the Phylogenetic Constraints Hypothesis. In general, the details of life histories and population dynamics are not adequately known for an informed investigation of these kinds of species (cf. Hunter 1995b). Clearly, there are many opportunities to advance this field of an evolutionary approach to the distribution, abundance, and dynamics of insect behavior.

Hunter's (1991, 1995a, b) analyses of outbreaking and non-outbreaking macrolepidopteran moth species and the evolution of flightlessness in the group showed that most families contain species that can be ascribed to each category of pest status and flight capability. Why such discrepancies occur is unknown, but some broad patterns emerge.

> There are seven origins of reduced wings ... Spring feeding is closely associated with wing reduction ... Half of the wing-reduced species overwinter as eggs as compared to only 21% of macropterous species ... Reduced wings are 11 times more likely to occur in egg-wintering lines than lineages that overwinter in other stages, a significant association ... Five of the seven origins of reduced wings involve winter-active adults ... The

GEOGRAPHIC RANGE	Large		Small	
HABITAT SPECIFICITY	Wide	Narrow	Wide	Narrow
LOCAL POPULATION SIZE				
Large, dominant somewhere	Locally abundant over a large range in several habitats	Locally abundant over a large range in a specific habitat	Locally abundant in several habitats but restricted geographically	Locally abundant in a specific habitat but restricted geographically
Small, non-dominant	Constantly sparse over a large range and in several habitats	Constantly sparse in a specific habitat but over a large range	Constantly sparse and geographically restricted in several habitats	Constantly sparse and geographically restricted in a specific habitat

Fig. 7.10. Rabinowitz's (1981) classification of seven forms of rarity in plants. The only cell with common organisms is at the top left, with a species covering a large geographic range, with wide habitat specificity and large population size.

number of tree genera in the diet is higher for wing-reduced species than for macropterous species in the Geometridae...Most wing-reduced species cluster their eggs or place them in a single mass, whereas most macropterous species spread them singly. (Hunter 1995a, pp. 279–280)

A commonly held view on nonoutbreak species is that populations are regulated by natural enemies (e.g. Mason 1987). "A preponderance of evidence supports the view that natural enemies are a principal force in keeping populations of forest Lepidoptera at low densities" (p. 50). However, Mason admits that little is known about nonoutbreak defoliators, "and virtually nothing is known about their population dynamics... If we are ever to answer the difficult question about population and community stability, much more attention will have to be given in the future to studying the uncommon along with the common species" (p. 52). However, considering rare species, Gaston (1994) found it impossible to generalize about causes: "The causes of rarity are by and large idiosyncratic, and beyond the broadest of generalizations it is impossible to predict in advance the reasons why any one species is rare" (p. 134).

If we subscribe to Rabinowitz's (1981) view on rarity in plants we would have a means of classifying species on a gradient from common to very rare based on geographic range, habitat specificity, and local population size (Fig. 7.10). This would indeed be an interesting exercise on insect herbivores, for it would help to provide some order among the vast number of species that fall into some kind of middle ground between highly eruptive species and very patchy, latent species. The advantage of Rabinowitz's view on seven forms of rarity is that species can be categorized in a descriptive manner, without any preconceived notions on the reasons why. Then comparative studies could be undertaken on species falling into different cells to investigate the bottom-up, lateral, and top-down factors that mechanistically account for differences. The approach would also emphasize constraints on species and the different life history traits that contribute to the evolutionary basis for understanding distribution, abundance, and dynamics. Gaston (1994) lamented the paucity of comparative studies on related rare and common species (see also Kunin and Gaston 1993).

My research group over the years has concentrated on extreme cases of uncommon and common species and our comparative approach based on the Phylogenetic Constraints Hypothesis has advanced understanding in my opinion. The main advantage of our approach was to study a patchily distributed species, which turned out to show latent population dynamics. Therefore we had a new vantage point for viewing eruptive species, because the population dynamics were relatively simple, easily understood, and comparable over related species. Much more of this kind of research is desirable, but there are large gaps in our knowledge in need of close inspection.

NOTABLE POINTS

The contrast between the latent and eruptive species is extreme and consistent for many kinds of species. When females evolve to be highly selective in oviposition a chain of evolutionary and ecological effects results, eventually leading to latent population dynamics, coupled with patchy distributions and relatively low abundance over a landscape. When females show little discrimination in oviposition in relation to larval food quality, either in time or in space or often both, their larvae fend for themselves and evolve with more general feeding habits, which results in the potential for defoliation of host plants.

The divergence of moths and butterflies is interesting because they share similar oviporus structures for laying eggs on foliage, not into plant tissue, and yet the butterflies have become much more specific in their oviposition behavior and much more similar in habits to other species with latent population dynamics. The phylogenetic shift in the butterflies to diurnal flight, accompanied by the evolution of excellent vision (Shapiro 1981; Rutowski in press), may be the key to understanding highly discriminating oviposition behavior, with the heliconiine butterflies perhaps showing the extremes in visual acuity, searching ability, and decision-making (cf. Gilbert 1975, 1991; Spencer 1988).

We have observed that several different kinds of life history result in similar kinds of population dynamics. Common themes run through different life histories. We see passive dispersal on a silken thread in the genus *Choristoneura*, in some macrolepidopterans such as the Douglas fir tussock moth and gypsy moth (Table 7.1), and in spider mites. Passive dispersal is also common in immature scale insects and adult aphids. Wingless adult females have evolved in moths, stick insects, aphids, and scale insects (and, of course, mites are wingless). And phase changes have evolved independently in three insect orders: stick insects, grasshoppers, and armyworms. Hence we can start to group different kinds of species into categories with similar evolved traits, either phylogenetic constraints or in the adaptive syndrome characters. As Lawton (1992) said, "There are not 10 million kinds of population dynamics." The simplest way to digest the diversity of species and dynamics down into a relatively small number of kinds is to use a phylogenetic approach. Another common theme in the eruptive species discussed in this chapter is females that oviposit independently of larval food quality. We might then expect larvae to evolve with characters akin to precocial animals. They forage for themselves from the time they hatch from the egg. In species with strong ovipositional preferences the larvae can evolve with more altricial-like characters, with mother essentially finding food for progeny, provisioning for larvae for much or all of their lives (cf. Fig. 2.11).

Lawton (1992) was thinking of classifying species more in terms of dynamical categories such as stable equilibria, limit cycles, chaos or random fluctuations. An admirable goal, indeed, but one that will require orders of magnitude more research effort to cover an adequate number of species and to research their dynamics in detail. And the will to do this is not now present, nor will it be ever present because the motivation to study the dynamics of many uncommon or nonpest species is lacking. However, taking an evolutionary approach and using the Phylogenetic Constraints Hypothesis opens an alternative view of nature and alternative means for classifying dynamical types. And each species does not need to be studied over many years to discover its dynamical properties. For example, we know that the mopane worm in southern Africa is eruptive, reaching very high numbers, and we know a little about its life history. And by comparison with species with similar phylogenetic constraints we can develop a mechanistic hypothesis on the dynamics and place the species in a large category of eruptive species.

We should also note, as we did for species with latent population dynamics, that the Phylogenetic Constraints Hypothesis is equally
valid and predictive in many vegetation types and at any latitude,
tropical or temperate. This is a compelling argument for a very strong
phylogenetic influence on ecology that in most cases trumps variation in ecological settings. An interesting exception is described in the
next chapter, but in general life history and morphological traits are
conserved in lineages, and as members of these clades move around
the world they act in a similar manner as phylogenetic constraints
and adaptive syndromes. I spent five most enjoyable and instructive
months in Brazil while on sabbatical leave in 1993 and 1994, going
with the specific expectation that in the tropics the Phylogenetic Constraints Hypothesis would not hold. I expected to find many exceptions, and these would aid in broadening the scope of my approach.
In fact, I found conformity, both to the Latent Species Hypothesis and
the Eruptive Species Model. I studied gall-inducing insects and membracids and herbivores attacking plants after fire in the cerrado, all
discussed in Chapter 6. I learned of agricultural pests that included
acridid grasshoppers, stick insects, cercopids that oviposit in the soil,
and pyralid moths, all discussed in this chapter. The conformity of tropical species to the Phylogenetic Constraints Hypothesis was striking. It
indicated that the predictive power of the hypothesis was enormous,
with much greater potential for extrapolation than the vast majority
of ecological hypotheses.

The pluralistic nature of the Phylogenetic Constraints Hypothesis is now fully apparent. It encompasses species with very different
evolved traits and emergent properties and accounts mechanistically
for their differences. It also is capable of grouping dissimilar kinds of
species into one category, such as eruptive species even though their
"starting-points" or phylogenetic origins are disparate. Consider acridid grasshoppers, forest lepidopterans, aphids, mites, scales, and stick
insects, all convergent in their potential for eruptive dynamics. This
Darwinian approach to the distribution, abundance, and population
dynamics of insect herbivores is also highly predictive, even though evo-
lutionists have apologized for or explained away the lack of prediction
in evolutionary theory (e.g. Mayr 1961). Given a certain kind of life history, we can predict the associated emergent properties. Given a certain
kind of dynamics, we can predict some aspects of a species' life history.

The contrast between the historical development of the field of
distribution, abundance, and population dynamics (Chapter 2), and the
Phylogenetic Constraints Hypothesis approach represents a paradigm

shift. From ecology to evolution, the shift is profound. Even when we consider the trends toward a more evolutionary view, discussed in Chapter 2, there is a radical move from a search for correlated traits of outbreak species to a mechanistic explanation of effects from evolved characters to ecological pattern. Certainly, Gilbert's (1975) view on *Heliconius* butterfly ecology being driven by evolved behavioral traits (Fig. 2.7) antedates by many years our first publication on phylogenetic constraints (Price *et al.* 1990). But, I believe that the *Heliconius* story was viewed very much as a remarkable and fascinating case of extremely refined coevolution. Not the kind of study from which we can derive broad generalizations. Time will tell whether the new paradigm will replace the old.

Weighing the evidence consistent with the phylogenetic constraints hypothesis in this chapter is largely for the future. However, the weight of evidence for outbreak lepidopterans, acridid grasshoppers, aphids, and scales as examples appears to favor the hypothesis that from an evolutionary point of view a large number of species should be expected to be eruptive. Likewise, those groups considered as contrasts to eruptive dynamics, the butterflies and bush katydids especially, lend credence to the hypothesis as it relates to species with latent population dynamics. Exceptions, of course, are numerous, but why they exist remains for future investigations to explain. Nevertheless, the hypothesis appears to be broadly applicable to insect herbivores, which should prompt us to ask if it is relevant to other taxa such as plants and vertebrates. I explore possibilities in Chapter 9.

8

Common constraints and divergent emergent properties

A serious potential weakness of the Phylogenetic Constraints Hypothesis is that the pine sawflies (Hymenoptera: Diprionidae), commonly severe pests and a sister group to the common sawflies (Tenthredinidae), show the same constraint of a sawlike ovipositor. This caused Berryman (1997) to reject the hypothesis because he argued that phylogenetic constraints in common between diprionid and tenthredinid sawflies should result in the same emergent properties. This may prove to be a common conception which needs to be addressed in this chapter. However, while the constraint may be the same in two or more families, the adaptive syndromes may be different, resulting in divergent population dynamics (cf. Price and Carr 2000). First, I will describe the differences between the families and then I will provide a hypothesis on why patterns in distribution, abundance, and population dynamics have diverged in these two sawfly families.

DIFFERENCES BETWEEN SAWFLY FAMILIES

One fundamental difference between tenthredinids and diprionids relevant to population dynamics is that a small percentage of tenthredinids are pest species (about 3 percent), whereas a large proportion of diprionids are serious pests in coniferous forests (about 40 percent in North America and 53 percent in Europe (cf. Larsson *et al.* (1993); for the largest genus in North America, *Neodiprion*, with 35 species, Arnett (1993) states that most are of economic importance). Many other differences in the families are evident (Table 8.1). Large tracts of conifer forest may be repeatedly defoliated, and trees eventually killed by diprionid sawflies. For example, the Swaine jack pine sawfly, *Neodiprion swainei*, killed large areas of jack pine, *Pinus banksiana*, in Ontario and Quebec provinces of Canada (Fig. 8.1). In monocultures of jack pine practically

Table 8.1. *Comparison of tenthredinids and diprionids in North America[a]*

Trait	Tenthredinids[b]	Diprionids
1. Number of species in North America	824	48
2. Number of pest/outbreak species	28	19
3. Percentage pests/outbreak species	3	40
4. Morphology	Females, light build Males, filiform antennae	Females, heavy build Males, pectinate antennae
5. Food plants	99 percent angiosperms	Exclusively conifers
6. Oviposition behavior	Commonly single eggs in carefully selected shoots	Often in egg clusters
7. Oviposition site	Exactly where larvae start to feed	On new foliage, but larvae feed on old foliage, or on old foliage depending on species
8. Maturity of plant tissue when eggs inserted	Very young growth	Young or old needles
9. Preference–performance linkage	Strong	Weak or absent
10. Flight	Moderately good	Poor, heavy with eggs
11. Egg synthesis	Synovigenic	Proovigenic
12. Fecundity	30–50 eggs in galling spp.	35–218 eggs
13. Larval feeding	Commonly solitary	Frequently gregarious

[a] Based on many sources and personal observations.
[b] Based on our studies discussed in Chapters 3 to 5.

every tree was killed in an outbreak over many kilometers (McLeod 1970, 1972) (Fig. 8.2). These monocultures develop naturally on poor sandy soils, and the species is fire adapted so that a jack pine stand is likely to be replaced by a new generation of jack pine after fire. Under natural conditions I cannot think of a single species of

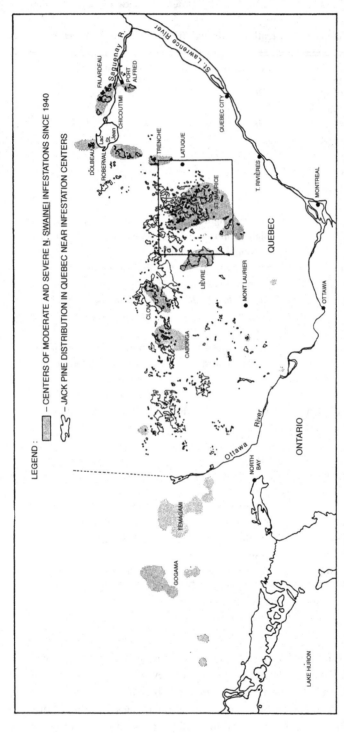

Fig. 8.1. Map of Ontario and Quebec provinces, Canada, showing "hazard areas" where outbreaks of Swaine jack pine sawfly recurred over 30 years up to 1967. Hazard areas are shaded and jack pine distribution is outlined. The boxed area is about 48 kilometers (30 miles) wide and the infestation in the St. Maurice River Valley is about 20 kilometers (12 miles) wide and twice as long. The Lake Oriskany outbreak occurred within this drainage, as detailed in Fig. 8.2. (From McLeod 1970.)

tenthredinid sawfly on angiosperms that causes such devastation. Some tenthredinids have evolved to attack conifers, and on these hosts they can cause serious damage, so we need to consider these kinds of species later in this chapter.

Part of the divergence in population dynamics from typically latent dynamics to commonly eruptive populations can be explained by differences in life history traits and female behavior: divergence in the adaptive syndromes of tenthredinid and diprionid sawflies. Female tenthredinids are generally lightly built relative to diprionids; wing loading is relatively low, so that flight is moderately good (Fig. 8.3). This enables a female to search in detail among host plants and within host plants for suitable oviposition sites. Tenthredinid females are synovigenic; they mature eggs gradually during the life of the female, they carry few mature eggs at any one time, and they appear relatively slim, aiding active, maneuverable flight. "The eggs are usually deposited singly though a number of oviposition incisions may be made in a single leaf" (Gauld and Bolton 1988, pp. 116–117). My own observations on diprionid females indicate a very different flight ability. Females are proovigenic and laden with eggs when they emerge, appearing rotund and heavy, with relatively high wing loading (Fig. 8.4). They are cumbersome flyers as a result, especially when laden with 100–200 eggs. "Diprionids are rather slow-flying, clumsy insects" (Gauld and Bolton 1988, p. 115). Their difficulty in gaining height during flight reminds one of small bombers struggling to take off. When so laden, females are unable to search among trees or within trees for suitable oviposition sites. They struggle up to any shoot they can reach and commence laying eggs. Because all eggs are mature they oviposit in large groups on a shoot, many species using new, young needles to saw into, followed by egg insertion. When such egg clusters hatch, larvae are aggregated and frequently remain together, united in colonial feeding. As with the flightless moths in which colonial feeding is common, the trait is derived from females with limited maneuverability, and is simply a trait correlated with outbreak dynamics rather than a causative agent in dynamics.

Tenthredinid females, with few eggs to lay at any one time, relatively good flying ability, and probably relatively low fecundity (50 eggs maximum in gall-inducing sawflies), spend much time searching for oviposition sites. They certainly select carefully in the species we have studied, demonstrating a strong ovipositional preference and larval performance linkage (cf. Chapter 3). Selection of long, vigorous shoots, which exist at low densities in typical willow stands, results in a low

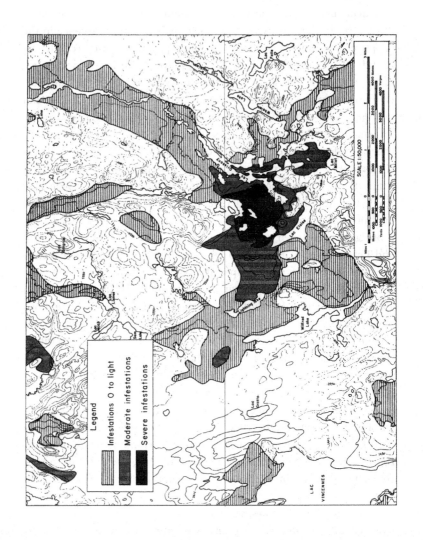

Legend

Infestations 0 to light

Moderate infestations

Severe infestations

SCALE 1:50,000

LAC
VINCENNES

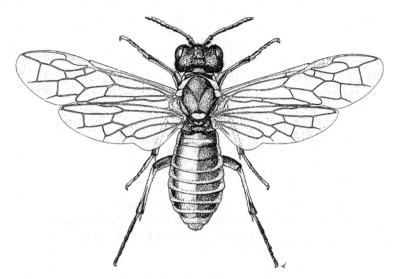

Fig. 8.3. An example of a female tenthredinid sawfly, *Tenthredo arcuata*, a member of the type genus. (From Gauld and Bolton 1988.)

carrying capacity for any sawfly population. Quite the reverse situation exists for diprionid sawflies. Females do not have the flight capacity to be selective; they fly ponderously and oviposit where they land, showing no oviposition preference except for the correct species of tree and young needles. Even then the oviposition sites are separate from feeding sites. In the species that oviposit in the spring on young needles, larvae feed on the previous year's and older needles. Young needles are too resinous for larval feeding. For diprionid species that oviposit into old needles in the fall, larvae do not hatch and begin feeding until the following spring, a separation of many months. In both cases, females without strong ovipositional preferences in relation to the quality of larval food will result in selection of larvae favoring those that can feed on any foliage they encounter: larvae become capable of feeding on all foliage on a tree or in a forest except for new foliage. Even this will be consumed later in the season when alternatives do not exist, contributing to the death of trees. The diprionid sawflies

Fig. 8.2. The Lake Oriskany outbreak of Swaine jack pine sawfly in the St. Maurice River Valley, Quebec Province, Canada, in 1965, showing areas of jack pine with severe, moderate, and light or no defoliation. The area of high sawfly densities was about 2.5 kilometers (1.5 miles) in width, in which virtually all trees were killed. The severe plus moderate infestations were twice as wide. (From McLeod 1970.)

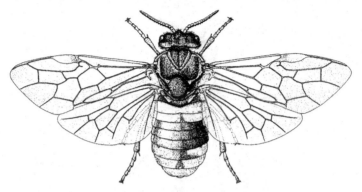

Fig. 8.4. *Diprion pini*, a member of the type genus of the Diprionidae. Note the thick-set, robust nature of the female depicted, with the "broad shouldered" appearance relative to *Tenthredo* in Fig. 8.3. (From Gauld and Bolton 1988.)

have retained high specificity to utilization of a single host species in many cases, as in the tenthredinids, but they have not evolved with the strong ovipositional preference within the host plant species. Hence, as with other outbreak species, they have the evolved capacity to be eruptive. Whether they actually demonstrate eruptive population dynamics depends on several different factors in their ecology: ecological factors limiting population growth such as natural enemies, weather, or shortage of food plants.

EVOLUTION OF PROOVIGENESIS IN DIPRIONIDS

The fundamental question to ask about the major difference in dynamics between tenthredinids and diprionids is, therefore, why tenthredinids evolved with synovigenesis and diprionids with proovigenesis. My hypothesis is that these different egg production strategies are limited to the different modes of growth in their food plants. Most angiosperms show indeterminate growth of shoots, resulting in active meristems, young leaves, and lengthening shoots over a prolonged part of the year, even in north temperate climates (cf. Niemalä and Haukioja 1982). Such growth patterns provide an extended window for oviposition into plant tissue (Fig. 8.5). Without serious limitations of time for ovipositing, a synovigenic strategy would appear to be favored, for females can devote time for finding high-quality sites. In contrast, conifers exemplify determinate growth, adapted to cooler climates with shorter growing seasons than for most angiosperms. Shoots and needles flush rapidly and

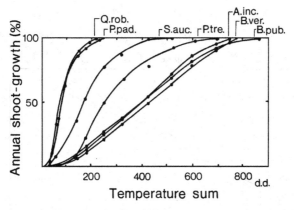

Fig. 8.5. Shoot growth of deciduous trees in Punkaharju Forest Research Station in southeastern Finland, based on studies by Raulo and Leikola (1974) and Niemelä and Haukioja (1982). Development is based on the "temperature sum," which is the cumulative sum of daily mean temperatures above 5 °C. The figure shows the brief availability of new foliage in species with mostly determinate growth (Q. rob. = *Quercus robur* and P. pad. = *Prunus padus*) and longer availability among other species that produce new leaves through much of the season (A. inc. = *Alnus incana*, B. ver. = *Betula verucosa*, and B. pub. = *Betula pubescens*). Intermediate species include *Sorbus aucuparia* (S. auc.) and *Populus tremula* (P. tre.). (From Niemelä, P. and E. Haukioja (1982) Seasonal patterns in species richness of herbivores, Fig. 2, *Ecol. Entomol.* 7: 169–175, Blackwell Science, Oxford.)

growth ends relatively early compared to angiosperms in the same locality. Thus, a short window of oviposition opportunity probably selects for proovigenesis for the species ovipositing into new needles (Fig. 8.6); I will call this the **Determinate Growth in Conifers Hypothesis** on sawfly dynamics. Pines show a growth pattern of shoot elongation that precedes needle elongation, making the time that needles are young and available for oviposition very brief relative to most angiosperm species (Fig. 8.6). An additional feature of conifers is the very low probability of shoot abscission, as seen on willow discussed in Chapter 3.

Niemelä and Haukioja (1982) have already suggested that angiosperm trees with more determinate growth, such as oaks, support fewer late-season insect herbivores than trees with indeterminate growth. This is because young leaves are available for much of the active season when growth is indeterminate, providing good quality food for herbivores for most of this time. It is also worth noting that species such as oak, with more determinate growth and a very short availability of

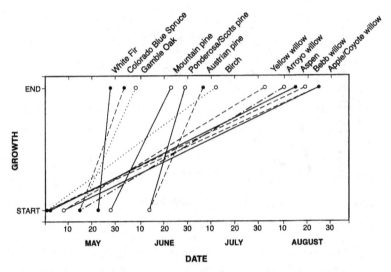

Fig. 8.6. Growth rates of new shoots and leaf and needle production in a common garden in Flagstaff, Arizona, showing the differences among conifer and hardwood trees. Dates of the first leaves per plant species and the last leaves suitable for oviposition by sawflies were recorded. Note the contrast between most angiosperms and conifers. Only Gamble oak flushes as rapidly as the conifers. Original observations.

young leaves (cf. Fig. 2.5), suffer from serious outbreak species such as winter moth and green oak tortrix moth (*Tortrix viridana*).

TENTHREDINIDS ON CONIFERS

Interesting, and relevant to this hypothesis on the divergence of oviposition strategies in tenthredinids and diprionids, is the presence of some tenthredinids that feed on conifers. A significant proportion of these are known to be serious pests in forests. Clearly, these tenthredinids have moved from lineages utilizing angiosperms. Only three genera of common sawflies in North America have species feeding on conifers: *Pristiphora*, *Pikonema*, and *Anoplonyx*. *Pristiphora* species feed mostly on angiosperms, including willows, oaks, birches, roses, and blueberries, 22 species in all, but three species feed on conifers: *Pristophora erichsonii*, the larch sawfly, a serious pest of *Larix laricina* in north temperate forests; *Pristophora leechi*, which feeds on *Larix occidentalis*, and *Pristophora lena* on *Picea* species. In the genus *Pikonema* only three species occur, all on spruces: *Pikonema alaskensis*, the yellow-headed spruce sawfly; *Pikonema dimmochii*, the green-headed spruce sawfly, and

Table 8.2. *Numbers of species of tenthredinids and diprionids in North America, with estimates of the percentage of outbreak or eruptive species*

	Number of species	Number of outbreak species	Percentage of outbreak species	Reference
Tenthredinidae on angiosperms	814	24	3	Smith 1979; Haack and Mattson 1993
Tenthredinidae on conifers	10	4	40	Smith 1979; Haack and Mattson 1993
Tenthredinidae, all species	824	28	3	
Diprionidae, all on conifers	48	19	40	Smith 1979; Larsson *et al.* 1993

Pikonema ruralis without a common name. *Pikonema alaskensis* is an outbreak species. Four species of *Anoplonyx* are described for North America, all feeding on larches, none of which are regarded as outbreak species.

A summary table of tenthredinids and diprionids shows the large discrepancies in outbreak dynamics and the trend in tenthredinids to become more eruptive on conifers (Table 8.2). Tenthredinids on angiosperms have a very low 3 percent of species that are regarded as outbreak species (cf. Haack and Mattson 1993), whereas diprionids are represented by 40 percent of species that are known to outbreak (cf. Larsson *et al.* 1993). But the few species of tenthredinids on conifers show, remarkably, an identical frequency of outbreak species when compared to diprionids.

This convergence in outbreak frequency of tenthredinids on conifers and diprionids, all on conifers, suggests that traits of conifers may result in selection for proovigenic egg development with consequent rapid, aggregated oviposition behavior, reducing the probability that females are selective of oviposition sites in relation to quality of larval food. Rapid, deterministic growth of conifer shoots and the short time that needles are available for oviposition appear to be sufficient to cause selection for the life history traits of diprionids. More diprionids oviposit into young foliage (32 species) than old foliage (14 species; cf. Larsson *et al.* 1993), placing more species under strong selection for rapid oviposition. Although rapid oviposition into old needles may also be selected for, the reasons are not clear. We would also like to know

Table 8.3. *Comparison of tenthredinids on conifers and diprionids*

	Tenthredinids		
Character	*Pristiphora*	*Pikonema*	Diprionids
1. Host plants	Conifer–larch	Conifer–spruce	Conifers
2. Adult emergence	Spring	Spring	Spring 70 percent spp., fall 30 percent spp.
3. Oviposition site	Shoots	Needles and shoots	Needles
4. Egg synthesis	Proovigenic?[a]	Synovigenic	Proovigenic
5. Module age utilized for oviposition	Current–new	Current–new	Current–new 70 percent spp., old 30 percent spp.
6. Fecundity	60–206	95	35–218
7. Egg cluster size	9–83 per shoot	?	1–38 per needle, 62–85 per shoot
8. Needle age fed on by larvae	New	New → old	Old and current
9. Larval feeding behavior	Gregarious	Gregarious	Gregarious 70 percent spp., Solitary 30 percent spp.
10. Number of years of feeding resulting in host plant death	Several	1 year	1 to several years

[a] Assumed, based on large numbers of eggs per shoot.
Sources: Tripp 1957; Drooz 1960; Houseweart and Kulman 1976; Haack and Mattson 1993; Larsson *et al.* 1993.

whether oviposition into young needles and old needles evolved only once or repeatedly and, if origins are monophyletic, which is the more primitive. Then scenarios for the evolution of life history traits and selection advantages could be developed. In the meantime, we are left to wonder.

In addition to similarities in percentage of outbreak species in tenthredinids on conifers and diprionids, many life history traits, behaviors, and host utilization patterns are similar (Table 8.3). Tenthredinids emerge in the spring, as do 70 percent of diprionids, and a similar pattern prevails for module age utilized, larval feeding behavior, and propensity for killing host trees. Fecundities are relatively high and clustering of eggs is evident and correlated with gregarious larval feeding. Larch is the only deciduous conifer utilized by sawflies, so only new foliage is available as larval food, whereas needles persist for several years in the other host species. I assume that the larch sawfly is proovigenic because of the many eggs laid into a single shoot and the high fecundity, as in diprionids, but *Pikonema* is clearly synovigenic (Houseweart and Kulman 1976).

DIFFERENT RESOURCES AND ADAPTIVE SYNDROMES

The evidence is largely consistent with the hypothesis that rapid, determinate growth in conifers results in selection for sawfly traits, which leads to the potential for eruptive population dynamics, in both diprionids and the few tenthredinids on conifers. Indeterminate host plant growth of many angiosperms seems to favor synovigenic egg development and the careful selection of oviposition sites by many tenthredinid species utilizing angiosperms.

Therefore, the nature of the resource utilized by insect herbivores influences their evolution significantly. First, conifers tend to grow in extensive monocultures, providing a high carrying capacity for insect herbivores. Second, shoot and needle growth selects for traits in sawflies likely to result in the potential for outbreak dynamics. Angiosperm trees also form extensive monocultures or mixed woodlands with few tree species present, for example aspen, oak, beech, and birch. But none to my knowledge is severely defoliated under natural conditions by sawflies. This is consistent with the determinate growth in conifers hypothesis.

In their extensive reviews on outbreak species of sawflies, Larsson *et al.* (1993) and Haack and Mattson (1993) considered the life history trait correlates of eruptive dynamics. Both Larsson *et al.* (1993) and Haack and Mattson (1993) detected gregarious larval feeding as a common character of outbreak species and discussed the advantages of gregarious feeding. In addition, Haack and Mattson (1993) showed that most outbreak-prone species had the following tendencies: to be more fecund, to initiate larval feeding early in the year, and to be more likely to cause tree death. However, as stated before in this book, the search for correlates of certain dynamic types leaves open the question of why such traits have evolved. I argue that the Phylogenetic Constraints Hypothesis coupled with selective factors involved with plant module development, as in the deterministic growth in conifers hypothesis, can account for the evolution of gregariousness. And I have argued that gregarious feeding results from selection for rapid oviposition by females, and the latter trait is critical to the evolution of the potential for eruptive dynamics. Gregarious feeding has evolved many times in insect herbivores and there are probably several selective scenarios involved, but it is a trait also found in many species that do not outbreak: "36% of the nonoutbreak sawflies are gregarious" (Haack and Mattson 1993, p. 534).

NOTABLE POINTS

Even with phylogenetic constraints in common, taxa can diverge in adaptive syndromes resulting in very different emergent properties. The divergence of the tenthredinids and diprionids appears to be associated with the utilization of angiosperms by most tenthredinids and conifers by all diprionids. The deterministic growth in conifers hypothesis is the first attempt at explaining why adaptive syndromes of the two taxa should diverge so significantly.

The convergence of conifer-feeding tenthredinids toward the diprionid adaptive syndrome and emergent properties lends credence to the hypothesis.

The pluralistic nature of the Phylogenetic Constraints Hypothesis is again evident, although for a different reason than discussed before on divergent constraints. Here we observe the same constraints but divergent kinds of resources for ovipositing females causing divergent adaptive syndromes; one results in latent population dynamics and one leads to eruptive dynamics.

That 60 percent of sawflies on conifers do not show outbreak dynamics needs more consideration, but this fact does not contradict the phylogenetic constraints hypothesis. The hypothesis states that the evolutionary potential for outbreak dynamics exists when females show little or no ovipositional preference. And the prediction is supported in 40 percent of the species. And the other 60 percent of species probably do have the potential for outbreak dynamics, yet to be revealed, probably because current environmental constraints act persistently and effectively. Alternatively, different criteria may reveal more species of economic importance as Arnett's (1993) estimates suggest.

9

The thesis applied to parasitoids, plants, and vertebrate taxa

This chapter is devoted to an examination of possibilities for extending the Phylogenetic Constraints Hypothesis to taxa other than insect herbivores. My ability to accomplish an insightful overview of any
other taxon is severely limited for a number of reasons. First, my major
research emphasis has been on insects and I know no other group as
well. And key observations, intuition, and "strong inference," as Platt
(1964) called it, all play a role in the development of concepts, in biology, usually based on familiarity with a particular organismal group.
Moving out of a realm of professional focus certainly lessens the credibility of interpretations developed. Second, insect herbivores live on
or in and feed on a very clearly defined resource, the living plant. We
can accurately evaluate this resource in many different ways and we
can examine precisely how insects respond to resource variation. All
of this is much more difficult with plants, granivorous birds, insectivorous birds, bats, and amphibians and planktivorous fish. Third, for
each new taxon I consider there is a vast literature characterized by
overwhelming detail and underwhelming synthesis, especially that relevant to the thesis proposed in this book. And even with the generous
advice of experts in each taxon, I feel that I am only tickling the surface
of the question on how the hypothesis applies to plants and vertebrate
taxa. A more authoritative treatment requires more space and more
time.

My choices of taxa to examine in this chapter are partially governed by a desire to move eventually through the trophic levels from
plants to herbivores to carnivores so that the major terrestrial trophic
levels of living organisms are integrated in a conceptual theme. This
path is not clear at present but some pointers and parallels developed from the foregoing chapters are worth consideration. The major impediment to progress on a broader front is the shortage or

absence of investigations on the population dynamics on a variety of plants and carnivores on herbivorous insects. After the plants I turn to vertebrates only to suggest a few avenues worthy of more detailed examination.

One group that I have some experience with is the parasitoids, whose larvae are truly parasitic on insect hosts, while adults are free-living wasps and flies, the females of which oviposit in, on, or near hosts of the larvae. I will treat these first, representing the carnivore trophic level, and then move on to plants and vertebrate taxa.

PARASITOIDS

There are four advantages to concentrating on the parasitoid families Ichneumonidae (Hymenoptera) and Tachinidae (Diptera). First, I am personally familiar with members of these families, having studied them for my doctoral thesis on the parasitoids of a diprionid sawfly. Second, eggs and larvae of the groups tend to live in or on the host insect for much of the remaining life of the host after attack, frequently emerging from an overwintering stage such as a larva in a cocoon. Third, the ovaries of members of these families are in a relatively primitive condition, such that the number of ovariole components per ovary correlates well with potential fecundity. Ovarioles are the production lines for eggs, with more ovarioles for species with higher fecundity. Fourth, host insects provide a clearly defined, readily quantified resource, much like plant resources. We have many life tables and survivorship curves for insect species that define the resource for parasitoids and how numbers change during tenure in the parasitic phase. The last three features make the following analysis possible.

Living in or on the host insect places any parasitoid species or group in a straitjacket of limitations that define the patterns in fecundity, behavior, and host suppression observed in these families. If oviposition occurs early in the life of larvae of the host and emergence of the adult is late, then the parasitoid eggs and larvae are exposed to all the mortality seen in the life of the host (Price 1973, 1974). High death rates in the host are compensated for by high fecundity made possible partly by ease of discovering relatively abundant hosts early in the survivorship curve (Fig. 9.1). Living with an active host for much of its larval life selected for oviposition into the host and internal parasitism. Hence, immunosuppression is necessary, accomplished in some species by an associated virus inoculated into the host during oviposition. These are the elements of the adaptive syndrome

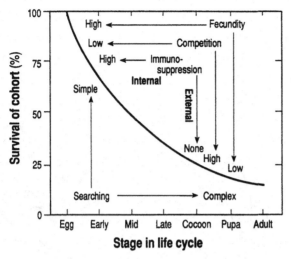

Fig. 9.1. Relationships among characteristics of parasitoids that attack insect hosts early and high on the survivorship curve, represented by the hollow curve, and those that attack low and late. These characteristics are associated as adaptive syndromes in response to phylogenetic constraints imposed by attacking living hosts with a parasitic way of life for parasitoid larvae. (From Price 1994a.)

for species and groups attacking high on the host survivorship curve.

Moving down the survivorship curve, hosts become less abundant than earlier, and as they enter overwintering sites hosts become better concealed. Searching for a host is therefore more complex and the rate of discovery declines and the time taken to oviposit into concealed hosts increases. Fecundity becomes relatively low in such parasitoid species but survivorship with the host has a much higher probability. Because hosts are relatively large and often fully grown, paralysis of the host, achieved by exact venom placement in thoracic ganglia in some cases, preserves the host for larval feeding. Larvae commonly feed externally, making immunosuppression unnecessary, increasing the probability of survival in a highly competitive niche with internal and external parasitoid species all present.

The patterns in fecundity for the families Ichneumonidae and Tachinidae show strong convergence in these insects from two different orders (Figs. 9.2 and 9.3) because of the severe limitations imposed by ovipositing into or onto living hosts generally with steep survivorship curves (cf. Cornell and Hawkins 1995; Price 1997). And the adaptive

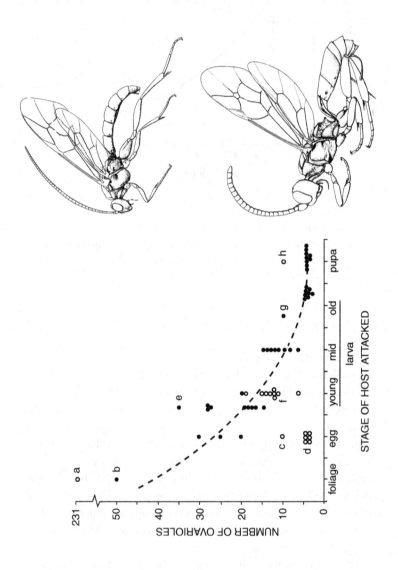

syndromes are clear enough relating to where in the life of a host oviposition occurs. Strong deviations from trends are observed when a major shift in strategy occurs. For example, ichneumonids ovipositing into eggs usually emerge from larvae and so conform to the general pattern, but some species oviposit into spider's egg cocoons, acting as egg predators. Thus, searching is difficult but probability of survival is high and fecundity has converged on the strategy of species attacking late larvae and pupae in cocoons (point d in Fig. 9.2). More accurately, those species attacking spider cocoons are close relatives of the cocoon parasitoids.

Each subfamily of the Ichneumonidae appears to be phylogenetically constrained to attacking a particular stage in the host's life cycle (Fig. 9.4). Hence, mean ovariole number per ovary in relation to host stage attacked, using means per subfamily, shows a similar pattern to that seen when individual species are used (Fig. 9.2). Clearly, the constraints and adaptive syndromes are complex enough to result in conservative adaptive radiation in relation to time of host attack. Of course, with so many potential hosts over which a radiation can spread, even the straitjacket of constraints discussed above does not prevent extraordinary speciation. Townes (1969), the reigning expert on ichneumonids in his time, estimated a world total of about 60 000 species in the family, more than all vertebrate species combined. When compared to about 9000 bird species assigned to their own class of animals, ichneumonids in only one family are almost seven times more speciose, with equal or greater structural diversity. "An ichneumonid genus is thus often equivalent to a bird family, and an ichneumonid subfamily to a bird order" (Townes 1969, p. 4).

Fig. 9.2. (Opposite) Relationships among number of ovarioles per ovary, which correlates well with total potential fecundity, in species of the parasitoid family Ichneumonidae, in relation to the host stage attacked. Note how egg production capacity declines in tune with the generalized survivorship curve of hosts. Exceptional points are noted as open circles and are discussed in detail in Price (1975). For example at point d, six species attacking eggs actually utilize the eggs of spiders in cocoons and are related to species that attack insect larvae in cocoons both phylogenetically and in the number of ovarioles per ovary. (From Price 1975.) Insets are a member of the genus *Euceros* (above) which lays eggs on foliage, with the highest number of ovarioles known in the Ichneumonidae (points a and b), and a member of the genus *Endasys* (below), which attacks larvae and pupae in cocoons and has only three ovarioles per ovary.

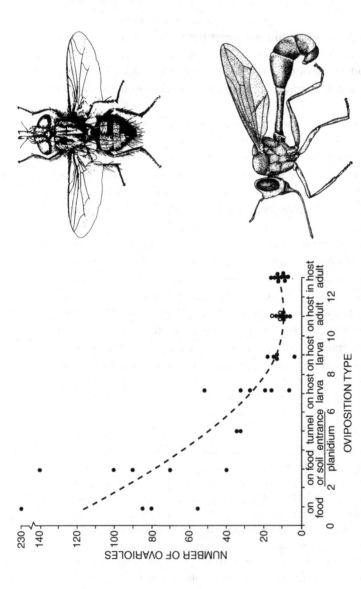

Fig. 9.3. Relationships among the number of ovarioles in parasitoid flies in relation to oviposition type and stage of host attacked. Solid circles are points of individual species of the Tachinidae and open circles represent species in the family Conopidae (Diptera). The planidium is an active first instar larva that can seek out concealed hosts, enter them, and then feed parasitically. Conopids oviposit into adult bees and wasps during flight and larvae develop endoparasitically. (From Price 1975.) Insets represent a tachinid (above) and conopid fly (below).

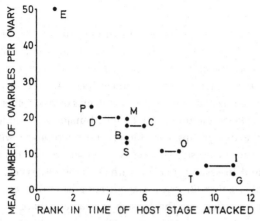

Fig. 9.4. The relationships between the stage of host development from egg to pupa and the number of ovarioles per ovary expressed as the mean of the specimens in the subfamily in the Ichneumonidae indicated by a letter. Details are in Price (1973). The rank 0–2 applies to species laying eggs on foliage, with an active planidial first instar larva that climbs onto a host insect. The following categories apply to attack on eggs (2–4), young larvae (4–6), up to pupal, puparium, or cocoon stage attacked (10–12). Note that subfamilies of ichneumonids are constrained to attack a narrow range of all possible stages in the life cycle of hosts. (From Price, P. W. (1973) Reproductive strategies of parasitoid wasps, Fig. 1, Am. Nat. 107: 684–693, University of Chicago Press.)

The case of the Ichneumonidae illustrates well the evident fact that when strict ecological limitations are imposed, through a mode of resource exploitation such as parasitism, accompanied by phylogenetic constraints of a piercing ovipositor and adaptive syndromes, adaptive radiation may be accelerated and species diversity increased dramatically. This is because the constraints limit the scope of radiation within one adaptive syndrome. Breakthroughs out of one set of limitations into a new set necessitate new adaptive syndromes and new radiations. Hence, heavy ecological limitations result in wave after wave of adaptive radiations when on resources as diverse as insect herbivore taxa.

If only we had good comparative data on parasitoid species distribution, abundance, and population dynamics, something could be said about the link between adaptive syndromes and emergent properties. But the motivation to understand dynamics of pest species has not been extended to the study of parasitoid populations except as they cause mortality of insect hosts. As is so frequently the case, we do not have the necessary information to test the Phylogenetic Constraints Hypothesis adequately.

PLANTS

The concepts developed in this book run through much of the literature on plant ecology and evolution over the long term, although the direct line of cause and effect from phylogenetic constraints to emergent properties is not evident. Raunkiaer's (1934) concept of life forms of plants recognized major adaptive strategies or syndromes that, while crossing phylogenetic lines, also included major taxonomic and phylogenetically related groups. For example, the therophytes with an annual life span, surviving stress as seeds, represent a major life history strategy found in several families with many weedy species. The spring ephemeral crucifers (Brassicaceae), *Bromus* grasses (Poaceae), and many *Chenopodium* species, like the goosefoots (Chenopodiaceae) all fit the therophyte mold. Another of Raunkiaer's groups is the geophytes, with buds underground on rhizomes, bulbs, or corms during the most stressful period of the year. Here again plant forms have strong family ties. The iris family (Iridaceae) contains species usually with rhizomes, corms, or bulbs. Many species in the families Liliaceae and Amaryllidaceae are also geophytes. The remainder of Raunkiaer's life form classes similarly have strong family ties: the hemicryptophytes represented by many grasses and rosette plants; chamaephytes illustrated by small shrubs; phanerophytes the trees and large shrubs.

Clearly, botanists have been interested in plant form for a long time because of the diversity in form, the obvious nature of differences, and the clearly adaptive features of form. *Plant form and function* was an early classic (Fritsch and Salisbury 1938) and Stebbins's (1950) *Variation and evolution in plants* relied very much on variation in form, as had so many of Darwin's botanical treatises: on orchids, climbing plants, and *The different forms of flowers on plants of the same species* (Darwin 1877).

Many features of plants are coordinated sets of traits, fitting the concept of adaptive syndromes perfectly. Indeed, the term "syndrome" is used explicitly for pollination syndromes, floral syndromes, and fruit and seed dispersal syndromes (e.g. Faegri and van der Pijl 1971; Real 1983; Howe and Westley 1986, 1988; Abrahamson 1989). And the nature of treating such syndromes is similar to that used so far in this book with insects as the subjects. For pollination syndromes, several are identified and flower traits for each syndrome are listed: time of opening, color, odor, shape, and nectar characteristics (Table 9.1). In my book, I have treated each adaptive syndrome separately, but they could be compiled into an equivalent table (cf. Figs. 2.8, 3.10, 7.2, 7.3, 7.6, 7.7, 7.9, and Table 8.1). In a similar way dispersal syndromes can be categorized (Table 9.2). Angevine and Chabot (1979) used the terms "germination

Table 9.1. *Pollination syndromes[a] that identify various taxonomic groups of flower visitors associated with certain flower design features*

Animal	Opening	Color	Odor	Shape	Nectar
				Flower	
Entomophilous					
Beetles	Day/night	Dull or white	Fruity or aminoid	Flat or bowl-shaped; radial symmetry	Often absent
Carrion or dung flies	Day/night	Brown or greenish	Fetid	Flat or deep; often traps; radial symmetry	Rich in amino acids, if present
Bee flies	Day/night	Variable	Variable	Moderately deep; radial symmetry	Hexose-rich
Bees	Day/night	Variable but not pure red	Sweet	Flat or broad tube; radial or bilateral symmetry	Sucrose-rich for long-tongued bees; hexose-rich for short-tongued bees
Hawkmoths	Night	White or pale green	Sweet	Deep, often with spur; radial symmetry	Ample and sucrose-rich
Butterflies	Day/night	Variable; often pink	Sweet	Deep or with spur; radial symmetry	Often sucrose-rich
Primarily pollinated by vertebrates					
Bats	Night	Drab, pale; often green	Musty	Flat "shaving brush" or deep tube; often on branch or trunk; hanging; much pollen; radial symmetry	Ample and hexose-rich
Birds	Day	Vivid; often red	None	Tube; often hanging; radial or bilateral symmetry	Ample and sucrose-rich

[a] "Pollination syndromes are suites of flower colors, scents, and shapes that serve as cues used by insect or vertebrate pollinators to locate flowers" (Howe and Westley 1988, p. 110). Such syndromes also maximize the probability of pollination itself.

Source: Howe and Westley (1988).

Table 9.2. Dispersal syndromes[a] and the kinds of animals that are favored by each type

Animal	Color	Odor	Fruit Form	Reward
Primarily dispersed by vertebrates				
Hoarding mammals	Brown	Weak or aromatic	Indehiscent thick-walled nuts	Seed itself
Hoarding birds	Green or brown	None	Rounded seeds or nuts	Seed itself
Arboreal mammals	Yellow, white, green, or brown	Aromatic	Arillate seeds or drupes; often compound and dehiscent	Pulp protein, sugar, or starch
Bats	Pale yellow or green	Musky	Various; often hanging	Pulp rich in lipid or starch
Terrestrial mammals	Often green or brown	None	Indehiscent nuts, pods, or capsules	Pulp rich in lipid or starch
Highly frugivorous birds	Black, blue, red, green, or purple	None	Large drupes or arillate seeds; often dehiscent; seeds >10 mm long	Pulp rich in lipid or starch
Partly frugivorous birds	Black, blue, red, orange, or white	None	Small or medium sized drupes, arillate seeds, or berries; seeds <10 mm long	Pulp often rich in sugar or starch
Feathers or fur	Undistinguished	None	Barbs, hooks, or sticky hairs	None
Primarily dispersed by insects				
Ants	Undistinguished	None to humans	Elaiosome on seed coat; seed <3 mm long	Oil or starch elaiosome with chemical attractant

[a] "Dispersal syndromes are constellations of fruit colors, scents, shapes, and nutritional qualities that are associated with different means of seed dissemination by biotic and abiotic agents" (Howe and Westley 1988, p. 119).

Source: Howe and Westley (1988).

syndromes" (p. 188) and "adaptive syndromes" (p. 194), adopting the latter term from their Cornellian colleagues soon after its original use by Root and Chaplin (1976).

Indeed, the whole idea of adaptive syndromes is conceptually easy to apply to plants and has been applied many times in various ways. To this extent the Phylogenetic Constraints Hypothesis applies readily to plants.

In other ways, at the macroevolutionary end of the hypothesis, the plant literature records the importance of evolved sets of characters such as in life histories, noting that various strategies can be defined (e.g. Harper and Ogden 1970; Gadgil and Solbrig 1972; Abrahamson and Gadgil 1973; Grime 1977; Menges and Kohfeldt 1995). As Schaal (1984, p. 188) stated: "Many life-history details underlie the basic patterns of reproduction and survivorship within a population." And many of the concerns raised in this book such as plant growth, phenology, and requirements for successful reproduction have been embraced as central to understanding aspects of plant ecology. For example, Grubb (1977), in his influential review on "the importance of the regeneration niche," recognized four component niches: habitat niche, life-form niche, phenological niche, and the regeneration niche.

Concerning constraints in the plant literature, especially phylogenetic constraints, there has been little emphasis. Harper (1982), following the initiative taken by Gould and Lewontin (1979), emphasized phylogenetic constraints, calling them "archetype effects." "No evolutionary process starts with a fresh sheet: always the process acts on ancestors that are more or less complex organized systems, and there are therefore limits on what new changes are possible...In each case the direction that an evolutionary pathway takes under selection will be under archetypic constraints" (Harper 1982, pp. 15–16). Certainly there has been an inherent recognition of phylogenetic constraints, as represented in the following argument by Niklas and O'Rourke (1982) and Tiffney and Niklas (1985) on the evolution of the earliest vascular plants.

> In general, all early land plants are assumed to have had a phototropic growth response due to selection among plants competing for light. However, continued production of vertical tissues by an apical meristem would eventually reach a point where the supporting tissues could no longer sustain the weight of the vertical axis. Failure of the vertical axis could...involve slow and continual deformation, leading to the gradual translocation of the products of vertical growth into the horizontal plane and thus to attainment of a rhizomatous habit...Thus the earliest

Table 9.3. *Defensive characteristics associated as adaptive syndromes for plants with rapidly expanding leaves and plants that expand young leaves slowly*

	Fast expanders	Slow expanders
Nitrogen content	High	Low
Toughness	Low	Medium
Chlorophyll	Low	High
Photosynthetic capacity	Low	Moderate
Chemical defenses	Low	High
Extrafloral nectaries	Common	Common
Synchrony	Common	Rare

Source: Coley and Kursar (1996).

vascular land plants were probably constrained by photosynthesis and mechanics to a clonal growth habit. (Tiffney and Niklas 1985, pp. 42–43)

So the phylogenetic constraint was recognized, but the adaptive syndrome included all the advantages of clonal growth (cf. Jackson *et al.* 1985).

As with this view on early vascular plants, views on constraints are frequently phrased in terms of physiological limits. A plant cannot grow rapidly and develop high chemical defense (Herms and Mattson 1992, 1994), or plants in the understory of a tropical rain forest are limited by light and nutrients (cf. Mulkey *et al.* 1996). Because plant ecophysiology is a strongly experimental science, phylogenetic issues rarely enter into the debate, but of course they are there. Taxa adapted to the exploitation of light gaps abound, such as the genus *Cecropia*, with accompanying syndromes for germination, pollination, and seed dispersal (cf. Brokaw 1985). And while Coley and Kursar (1996) emphasize physiological constraints on plant defenses against herbivores, they recognize familial patterns in how constraints and tradeoffs are resolved. For example, in their most effective long-term studies on adaptations for avoiding herbivory on young leaves in tropical forest plants they note, on one syndrome of rapidly expanding leaves followed by rapid toughening and chemical defense, that some families "consistently have rapid expansion, such as Connaraceae, Sapindaceae, and to a lesser extent Caesalpiniaceae" (Coley and Kursar 1996, p. 327) and they tabulate the characteristics of each (Table 9.3). Constraints and adaptive syndromes were clearly integrated in their work.

Any allusion to concepts akin to emergent properties in plant population dynamics has failed to grip my attention. Even the subject of plant population dynamics has not entered into the general debate on the topic of causes of population dynamics. Understandably so. Many plants have life spans beyond that of human investigators, and certainly beyond the normal limits of research grants. The others are opportunistic annuals or biennials, the ruderal species (Grime 1977, 1979), with interesting patch dynamics (e.g. Pickett and White 1985), but needing spatial models on dynamics rather than time-series models used in the study of insects and other animals. And the appropriate metapopulation models (e.g. Gilpin and Hanski 1991; Hanski and Gilpin 1997) have not been applied adequately to plants, although excellent examples of metapopulation structure exist (e.g. Quintana-Ascencio and Menges 1996; Quintana-Ascencio *et al.* 1998). In addition, plants frequently grow in complex mixtures, forcing a preoccupation with competition, and ecological succession.

However, the emergent properties in plant populations are clearly tangible and could be readily documented. With well-understood adaptive syndromes in plant form, germination, pollination, seed dispersal, and life history, the evolutionary basis for predicting population dynamics is well understood. But, perhaps it is too obvious to be interesting, heuristic, or revealing. When Grime (1977, 1979) identified three major strategies of plants, or adaptive syndromes, and listed the characteristics of plants with the competitive, stress tolerant, and ruderal strategies (Table 9.4), he could have added readily the expected kinds of population dynamics or emergent properties had he been inclined. He could also have added the plant families or genera that fitted the three syndromes and the phylogenetic constraints relevant to each.

My conclusion is that the application of the Phylogenetic Constraints Hypothesis to plants is straightforward and should be useful. More emphasis on the strong linkage between phylogenetic constraints, adaptive syndromes, and emergent properties would certainly advance the conceptual framework of plant ecology. It would aid in the synthesis of evolution and ecology. And the hypothesis should foster a more pattern-seeking, generality-finding emphasis. Even in 1986, Watkinson in his chapter on plant population dynamics, noted: "Despite the pioneering work of Watt (1947), population biologists are only just beginning to collect significant data on the spatial and temporal dynamics of plant populations" (p. 184), while noting that the factors that determine pattern in abundance and dynamics are in an early exploratory phase.

Table 9.4. *Grime's (1977) classification of three major plant strategies, competitive, stress-tolerant, and ruderal plants, and the characteristics of each group*[a]

	Competitive	Stress tolerant	Ruderal
Morphology of shoot	High dense canopy of leaves; extensive lateral spread above and below ground	Extremely wide range of growth forms	Small stature, limited lateral spread
Leaf form	Robust, often mesomorphic	Often small or leathery, or needle-like	Various, often mesomorphic
Litter	Copious, often persistent	Sparse, sometimes persistent	Sparse, not usualy persistent
Maximum potential relative growth rate	Rapid	Slow	Rapid
Life forms	Perennial herbs, shrubs, and trees	Lichens, perennial herbs, shrubs, and trees (often very long lived)	Annual herbs
Longevity of leaves	Relatively short	Long	Short
Phenology of leaf production	Well-defined peaks of leaf production coinciding with period(s) of maximum potential productivity	Evergreens with various patterns of leaf production	Short period of leaf production in period of high potential productivity
Phenology of flowering	Flowers produced after (or, more rarely, before) periods of maximum potential productivity	No general relationship between time of flowering and season	Flowers produced at the end of temporarily favorable period
Proportion of annual production devoted to seeds	Small	Small	Large

[a] In his subsequent book (Grime 1979) he doubled the number of characteristics associated with each strategy under major headings of morphology, life-history, physiology, and miscellaneous. Characteristics of demography and population dynamics could be readily added.

Source: Grime, J. P. (1997) Evidence for the existence of three primary strategies in plants and its relevance to ecological and evolutionary theory, *Am. Nat.* 111: 1169–1194, University of Chicago Press.

PLANTS AND HIGHER TROPHIC LEVELS

Certainly a blending of the Phylogenetic Constraints Hypothesis relevant to plant hosts and insect herbivores would advance both fields of ecology dramatically. It is intuitively obvious that plant constraints, adaptive syndromes, distribution, abundance, and dynamics must influence the herbivores feeding on these plants. For the bottom-up factors that move through the trophic system are strong and overriding when all work together. Consider some of these influences discussed in Price (2002): (1) The plant as food for herbivores. (2) The plant as habitat for the majority of herbivores, the arthropods and their carnivorous relatives, parasitoids and predators. (3) The constitutive chemicals in plants. (4) The induced changes in plants caused by herbivory. (5) The physical traits of plants such as size, toughness, and trichomes. (6) Traits of plants that require evolutionary responses by herbivores, such as crypsis, phenological synchrony, and life history, morphological and behavioral adaptations. (7) Landscape and biogeographic variation in vegetation types and food web richness.

I have attempted to blend patterns of plant life-form and other adaptive traits with consequences at the second and third trophic levels, using ecological succession as an organizing gradient (Price 1991a, b, 1994b). Examples for early and late north temperate plant succession are provided in Figs. 9.5 and 9.6. But there are serious limitations to this approach, not the least being a shortage of broad-scale comparative studies and the loss of strong phylogenetic associations from plants to herbivores to carnivores, and from early successional species to late successional species. Nevertheless, the approach does provide a conceptual framework on which to build comparative studies and on which to develop a synthesis on multitrophic level interaction webs based on living plants.

Other approaches linking plant adaptive strategies to herbivore traits have been developed by Coley and associates (e.g. Coley et al. 1985; Coley and Aide 1991; Coley and Kursar 1996). Plants were classified as inherently fast-growing or slow-growing with growth characteristics of each listed, adaptive syndrome style, as well as defense characteristics against herbivores (Table 9.5). Fast growers were less well defended, were able to compensate for damage because of high resource availability, and were much preferred by herbivores in tropical, cold temperate, and Arctic vegetation (Coley et al. 1985). Of course, there are considerable phylogenetic associations in plants with fast and slow growth and low and high constitutive phytochemical defenses, but

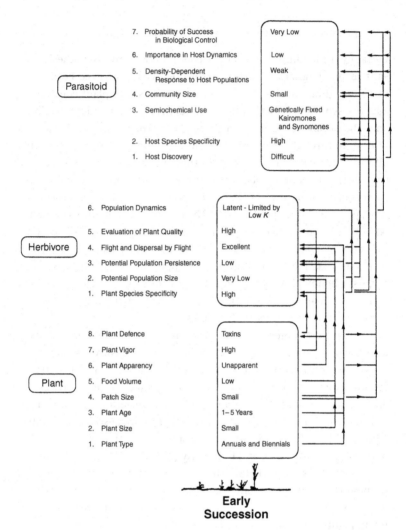

Fig. 9.5. Some traits of plants in early succession that influence insect herbivores directly and also parasitoids at the third trophic level. Linkages are justified in Price (1991a, b, 1994). (From Price 1994.)

these were not elaborated in the paper. And there are only limited phylogenetic parallels in plant and herbivore groups because herbivorous insects appear to be ecological opportunists shifting hosts and host taxa where habitat opportunities occur. Also, in tropical vegetation, there are the plant species with the rapidly-expanding-leaf syndrome and the slowly-expanding-leaf syndrome (Coley and Aide 1991; Coley and

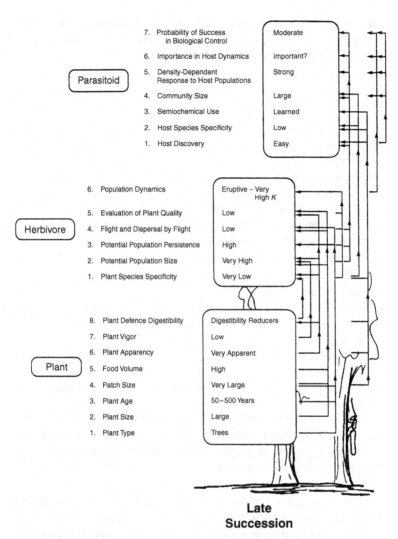

Late Succession

Fig. 9.6. In contrast to early successional plant traits listed in Fig. 9.5, the same sets of criteria are employed for late successional species of plants, insect herbivores, and parasitoids. Interacting effects up the trophic system are indicated. (From Price 1994.)

Kursar 1996) (see Table 9.3). Extrapolating to the effects on herbivores Coley and Kursar suggest the following:

> We speculate that the trade-off between expansion rate and chemical defense may have implications for herbivore population dynamics and life history traits. Species with rapidly expanding leaves may more easily

Table 9.5. *Characteristics of plants that grow relatively fast and slowly and some ecological conditions associated with these adaptive syndromes: resource availability and rates of herbivory*

Variable	Fast-growing species	Slow-growing species
Growth characteristics		
Resource availability in preferred habitat	High	Low
Maximum plant growth rates	High	Low
Maximum photosynthetic rates	High	Low
Dark respiration rates	High	Low
Leaf protein content	High	Low
Responses to pulses in resources	Flexible	Inflexible
Leaf lifetimes	Short	Long
Successional status	Often early	Often late
Antiherbivore characteristics		
Rates of herbivory	High	Low
Amount of defense metabolites	Low	High
Type of defense (*sensu* Feeny)	Qualitative (alkaloids)	Quantitave (tannins)
Turnover rate of defense	High	Low
Flexibility of defense expression	More flexible	Less flexible

Source: Reprinted with permission from Coley, P. D., J. P. Bryant, and F. S. Chapin (1985) Resource availability and plant antiherbivore defense, Table 1, *Science* 230: 895–899, © 1985 American Association for the Advancement of Science.

escape detection by specialist herbivores, and instead, because of their higher palatability, be fed on primarily by generalists. In contrast, slow expanders, with their variety of chemical defenses, would more likely be targets of specialists that are capable of handling the secondary metabolites. Third, higher levels of chemical defense in slow expanders may mean that larval development of herbivores is slowed. This prolonged period of larval vulnerability opens the possibility for predators and parasitoids to have greater control of herbivore populations. Increased pressure from the third trophic level may, in turn, dampen herbivore outbreaks. (Coley and Kursar 1996, p. 323)

We see in the plant literature the same kinds of emphasis as I have used in this book on insects, and a broad recognition of syndromes of integrated adaptive traits involving morphology, life history, physiology, and we can add plant behavior if we allow that growth patterns, say of leaf expansion rates, qualify (cf. Tables 9.3 and 9.4). If we had to select one major factor as a key constraint on plant adaptive syndromes, and how plants can most readily be integrated with the Phylogenetic Constraints Hypothesis, I think we might agree on plant size. Such a huge range in plant size provides a strong gradient on which to search for pattern and mechanistic explanations of pattern. A strong phylogenetic component is involved with size, and life history traits such as longevity, time of first reproduction, and mating system map onto plant size as well. Distribution, abundance, and no doubt population dynamics also relate well to size, or would do so if the data were available. In general, then, I see no barriers or impediments in extending the Phylogenetic Constraints Hypothesis to plants.

VERTEBRATE ANIMALS

As with plants, body size of vertebrates is a trait that is associated with broad macroevolutionary and macroecological patterns within and among taxa. And the mechanistic explanations for such patterns are well developed (e.g. Peters 1983; Calder 1984). At a very general level, then, I would agree that body size acts as a phylogenetic constraint and many attributes of animals are associated with body size. In the mammals, birds, reptiles, amphibians, and fish, we do not have major differences in design of apparatus for oviposition as in the insects that help us to identify major divergences in adaptive strategies. But there are major differences in size. Whether size is the basic constraint, or a trait that must vary in unison with many other characteristics, is certainly debatable. But body size is certainly the most tangible trait that shows trends in evolutionary time and ecological space, such as latitude; it is the best-studied characteristic of animals and the best understood.

Certainly, there are phylogenetic constraints associated with body size. Small animals have an adaptive syndrome of traits such as short generation time and life span, high growth rate, and high reproductive rate (Bonner 1965). The evident emergent properties are high population density, dramatic changes in density, and plague-like conditions in at least some species. Lemming, vole, mouse, and rat outbreaks are well documented, and are of special concern when grain harvests are threatened and the epidemiology of human diseases is involved. There are

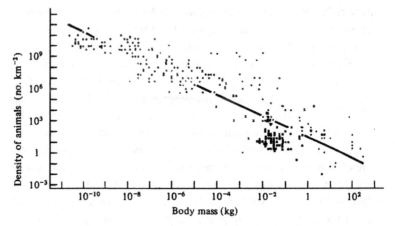

Fig. 9.7. The general relationship between body mass and density of animals, covering mammals, birds, reptiles, amphibians, and invertebrates. (From Peters, R. H. (1983) *The ecological implications of body size,* Fig. 10.2, Cambridge University Press.)

very general negative relationships among body mass and animal density which apply to birds, mammals, reptiles, amphibians, fish, and invertebrates (Peters 1983) (Figs. 9.7 and 9.8).

There are also phylogenetic patterns in body size, speciation, extinction, adaptive radiation, and the eventual taxonomic diversity of lineages. Small size and short generation time are the starting-points for a series of effects resulting in high taxonomic diversity (Marzluff and Dial 1991) (Fig. 9.9). The patterns are relevant to mammals, birds, amphibians, and reptiles. As Marzluff and Dial point out:

> Populations with high growth rates, especially very mobile ones, are expected to be good colonists and therefore often encounter new resources...Rapid turnover of individuals in the population can lead to strong selection and rapid evolution...Founding populations may therefore speciate rapidly as they adapt to new resources. Similarly, they may withstand extinction in the face of environmental deterioration because they can adapt to environmental changes and quickly attain large population sizes over broad geographic ranges...High mobility and rapid population growth rates also reduce extinction by enabling populations to recolonize areas in their range that experienced local extinctions...Short generation time also increases the rate of recombination and mutation...which may provide the raw material for the evolution of a unique key character commonly associated with adaptive breakthroughs. (Marzluff and Dial 1991, pp. 433–434)

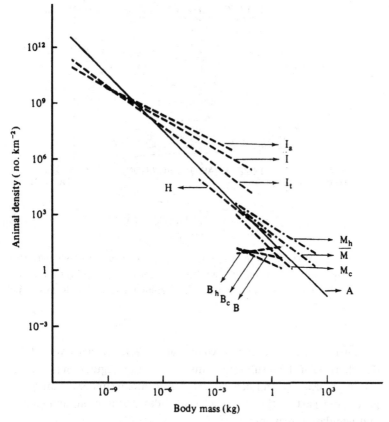

Fig. 9.8. The body mass and density of animals relationship for specific groups of animals as well as for all animals. A, all animals; B, birds; B_h, herbivorous birds; B_c, carnivorous birds; H, vertebrate poikilotherms; I, invertebrates; I_a, aquatic invertebrates; I_t, terrestrial invertebrates; M, mammals; M_h, herbivorous mammals; M_c, carnivorous mammals. (From Peters, R. H. (1983) *The ecological implications of body size*, Fig. 10.3, Cambridge University Press.)

The very apparent and inextricable link between evolutionary and ecological processes and traits are clearly illustrated in this passage. So much so that we must wonder at the separation of the two subjects in so many studies on population dynamics. However, as broader synthesis is sought, a strong evolutionary basis for understanding patterns in ecology is inevitable. For example, rodents are by far the most diverse group of mammals, and are at the small end of the range in body size of mammals, although not the smallest (Dial and Marzluff 1988). Most rodents

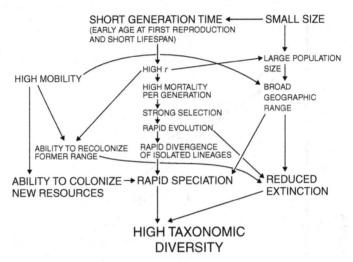

Fig. 9.9. A summary model provided by Marzluff and Dial (1991), starting with short generation time and small size in animals and resulting in consequences in population size, geographic range, and high taxonomic diversity.

are altricial, with nonmobile young, and are territorial (Wolff 1997) (Fig. 2.11), with the resultant intrinsic population regulation (Fig. 2.12). Evolution, ecology, and behavior all blend into an understanding of a group and part of that knowledge includes distribution, abundance, and population dynamics.

Concerning birds, Ricklefs *et al.* (1998, p. 282) stated that "The issue of constraint is central to understanding evolutionary diversification of life-history patterns." Thus understanding constraints is basic to elucidating the evolved traits that could be assembled as the adaptive syndrome in a clade of bird species. But the link among constraints, adaptive syndromes, and emergent properties is not well developed in the avian literature, even though much detailed information is known on life history, growth, and development (e.g. Starck and Ricklefs 1998). Promislow (1996) created a matrix of behavioral traits and population-level phenomena (Fig. 9.10), showing where comparative approaches had explored interactions, but also gaps in information especially in population density and dynamics. Nevertheless, he argued (pp. 303–304) that, "Although population dynamic traits may not follow directly from properties of the individuals within a population, we need not view these traits as emergent properties whose causal factors are inexplicable." He went on to explain for the first

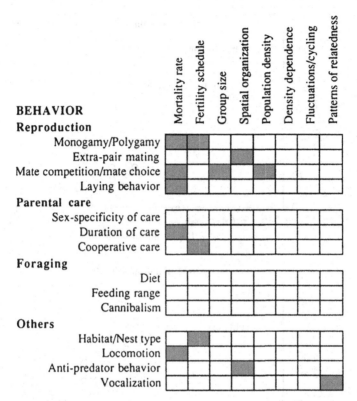

Fig. 9.10. Promislow's (1996) matrix of behavioral traits and population phenomena with shaded boxes showing areas in which comparative approaches have been investigated.

time variability in population size among bird species in Britain using adult survival as the independent variable (Fig. 9.11). Ricklefs (2000a) also showed strong patterns in empirical relationships among life table variables in birds. When density of bird species is plotted against body mass a negative relationship is observed repeatedly (Brown and Maurer 1987, 1989; Brown 1995) (Figs. 9.7 and 9.8). Body size and correlated life history and behavioral traits obviously have strong effects on population dynamics. "I have hypothesized that the morphology, physiology, and behavior of individual organisms play major roles in causing, or at least constraining, large-scale patterns of distribution and abundance, both within and among species" (Brown 1995, p. 119).

As with the literature on parasitoids and plants, the literature on vertebrates is compatible with the Phylogenetic Constraints Hypothesis,

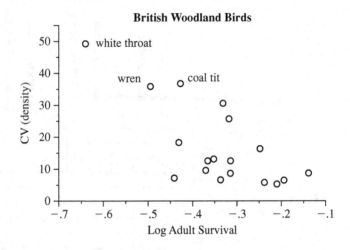

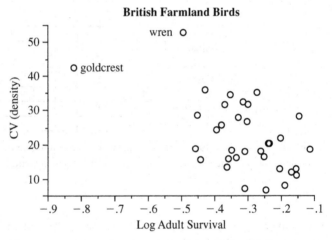

Fig. 9.11. Promislow's (1996) analysis of the relationship between the coefficient of variation (CV) in annual population density in wild populations of British birds and adult survival: British woodland birds (above) and farmland birds (below). CV values are from Pimm (1984). In both cases relationships are significant and negative.

but application to the understanding of distribution, abundance, and population dynamics is limited by the shortage of studies in population dynamics across the range of taxa in any one group. Comparative studies are in short supply. The sheer diversity of insect life in form, function, and life history and the pest status of so many species has resulted in a literature amenable to broad comparative work on population dynamics unmatched in other taxonomic groups. However, more holistic views on the population dynamics of many species will be beneficial in any kind of management practice from pest management, to conservation, vector relationships in epidemiology, and the treatment of invasive species. In these contexts the Phylogenetic Constraints Hypothesis offers a coherent and holistic conceptual approach to population management.

10

Theory development and synthesis

THE ECOLOGICAL VIEW OF POPULATION DYNAMICS

We should ask the question of what generalities are available from the Microecological Idiosyncratic Paradigm compared to the Macroevolutionary Nomothetic Paradigm. I am not equipped to make this comparison objectively because I have stated my views clearly enough in this book. But to foster debate, I would like to offer the following arguments based on population dynamics studies on insect herbivores of the longer kind, 5 years or more, or at least five generations. I have compiled a representative set of examples, with 62 species included, covering a broad spectrum of taxa (Table 10.1). This is certainly not an exhaustive list but is derived largely from several sources that have reviewed an aspect of insect herbivore population dynamics (e.g. Dempster 1983; Berryman 1988, 1999; Myers 1988; Watt *et al.* 1990; the papers edited by Liebhold and Kamata 2000) plus reprints in my own collection. Forest-dwelling species are well represented, reflecting what is probably a real bias in the literature because forest habitats are relatively stable with even individual trees persisting through a long study, disturbance is minor, foliage feeders and gallers are relatively easy to sample, and only rarely is the resource base – the trees – destroyed by the feeding (cf. Liebhold and Kamata 2000). The studies represent many different approaches in terms of sampling methods, surveys, records of damage, plot studies, and landscape views, observational and experimental. Even when life table generation and analysis have been employed, some have emphasized the detection of density-dependent factors, which may regulate populations, and others use K-factor analysis which detects the factor best correlated with total K, the total generation loss, over the years of study. Most studies have not included experimental tests of hypotheses, leaving cause and effect uncertain, a point repeated many times but

often forgotten (e.g. Morris 1959; Krebs 1995; Royama 1996; Ruohomäki
et al. 2000). As Morris (1959, p. 587) stated, "A key factor was defined as
any mortality factor that has useful predictive value and no attempt has
been made to establish cause and effect, because single-factor data are
scarcely suitable for this purpose." In the table I have not included the
form of analysis or the nature of the data used, and encapsulating major factors diagnosed as important oversimplifies the conclusions and
provides the appearance of certainty where none exists in reality. Nevertheless, the examples used provide the flavor of the literature based
on studies over the past 200 years.

What notable points or generalities emerge from a study of
Table 10.1?

1. Very long-term studies of defoliation covering many sites yield
important clues on the basis of outbreaks, but without more detailed
studies they remain as hypotheses only. The remarkable long-term monitoring of outbreaks of pine-feeding herbivores in Germany reported by
Klimetzek (1990) from 1801 to 1988 yields only some possibilities on
the causes (see spp. 29, 42, 49, and 54 in Table 10.1). And the 50- and
60-year studies reported by Schwerdtfeger (1935, 1941; Varley 1949) leave
few clues on causes of outbreaks (spp. 42, 47, 54, and 55).

2. More detailed long-term studies may fail to discern a clear understanding of mechanisms in population dynamics. Of course, these
are hard to publish so probably only a small percentage emerge into
the literature. Wool's (in press) 20-year study showed how the large
banana gall aphid, Baizongia pistaciae, in Israel (sp. 9) was difficult to
understand. And even though Lawton and associates conducted numerous experiments on the herbivore fauna on bracken fern, explanation
of dynamics remained elusive (Lawton 2000). Of the 27 species present
at Skipwith Common, studied for 19 years, 16 species were abundant
enough to provide census information each year, plus two aggregates
of species. In Table 10.1 I have chosen some representatives from major
groups of herbivores (spp. 7, 15, 17, 20, 22, 27, and 28), and in each case
clues of effects are mentioned or important factors are unknown, but
in no case is the dynamics understood. The difficulties with studying
very low populations, sparsely distributed on an apparently superabun-
dant resource, are apparent enough. But such problems, which have
contributed to a general disregard for studying low-population species,
leave a vacuum for comparative approaches on common and uncommon or numerous and sparse insect species.

Table 10.1. *Studies of herbivorous insects lasting five or more years or generations, and conclusions on regulation or limitation or host conditions favoring high populations*

Insect species	Taxonomic affinity	Host plant(s)	Locality	Study length	Major factor	$r^2 \times 100$[a]	Reference
1. *Didymuria violescens*	Phasmatodea: Phasmatidae	*Eucalyptus* spp.	Australia	17 yr	Weather, bird predation		Readshaw 1965
2. *Drepanosiphum platanoides*	Homoptera: Aphididae	*Acer pseudoplatanus*	U.K.	19 yr	Intraspecific competition		Dixon 1990
3. *Eucallipterus tiliae*	Homoptera: Aphididae	*Tilia vulgaris*	U.K.	20 yr	Intraspecific competition		Dixon 1990
4. *Phyllaphis fagi*	Homoptera: Aphididae	*Fagus sylvatica*	U.K.	19 yr	Intraspecific competition		Dixon 1990
5. *Thelaxes dryophila*	Homoptera: Aphididae	*Quercus robur*	U.K.	20 yr	Intraspecific competition		Dixon 1990
6. *Myzocallis boerneri*	Homptera: Aphididae	*Quercus cerris*	U.K.	14 yr	Intraspecific competition		Dixon 1990
7. *Macrosiphum pteridicolens*	Homoptera: Aphididae	*Pteridium aquilinum*	U.K.	19 yr	Unknown		Lawton 2000
8. *Cinara pinea*	Homoptera: Aphididae	*Pinus sylvestris*	U.K.	9 yr	Host quality, predation		Kidd 1988
9. *Baizongia pistaciae*	Homoptera: Pemphigidae	*Pistacia palaestina*	Israel	20 yr	Unknown		Wool in press
10. *Cardiaspina albitextura*	Homoptera: Psyllidae	*Eucalyptus blakelyi*	Australia	6 yr	Intraspecific competition, cool weather, predation		Clark *et al.* 1967
		Eucalyptus spp.	Australia	8 yr	Food shortage, predation		Morgan and Taylor 1988
11. *Fiorinia externa*	Homoptera: Diaspididae	*Tsuga canadensis*	U.S.A.		Plant stress, competition, parasitoids		McClure 1988

	Species	Order: Family	Host plant	Country		Factor	%	Reference
12.	*Nuculaspis tsugae*	Homoptera: Diaspididae	*Tsuga canadensis*	U.S.A.		Plant stress, competition, parasitoids		McClure 1988
13.	*Aleurotrachelus jelenekii*	Homoptera: Aleurodidae	*Viburnum tinus*	U.K.	13 yr	Weather	17–59%	Southwood and Reader 1976
14.	*Neophilaenus lineatus*	Homoptera: Cercopidae	*Juncus, Festuca* spp.	U.K.	37 yr	Temperature	70–75%	Whittaker & Tribe 1998
15.	*Ditropis pteridis*	Homoptera: Delphacidae	*Pteridium aquilinum*	U.K.	19 yr	Unknown		Lawton 2000
16.	*Lygaeus equestris*	Hemiptera: Lygaeidae	*Vincetoxicum hirundinaria*	Sweden	18 yr	Weather, food quantity	62%	Solbreck 1995
17.	*Chirosia histricina*	Diptera: Anthomyiidae	*Pteridium aquilinum*	U.K.	19 yr	Frond quality?		Lawton 2000
18.	*Lasiomma melania*	Diptera: Anthomyiidae	*Larix* spp.	France	10 yr	Cone crop, food quantity	92%	Roques 1988
19.	*Phytobia betulae*	Diptera: Agromyzidae	*Betula pendula, B. pubescens*	Finland	47 and 65 yr	Tree age and variation	77%, 64%	Ylioja et al. 1999
20.	*Phytoliriomyza* spp.	Diptera: Agromyzidae	*Pteridium aquilinum*	U.K.	19 yr	Unknown		Lawton 2000
21.	*Taxomyia taxi*	Diptera: Cecidomyiidae	*Taxus baccata*	U.K.	28 yr	Parasitoids	8–52%	Redfern and Cameron 1998
22.	*Dasineura filicina*	Diptera: Cecidomyiidae	*Pteridium aquilinum*	U.K.	19 yr	Frond quality?		Lawton 2000
23.	*Epilachna niponica*	Coleoptera: Coccinellidae	*Cirsium kagamontanum*	Japan	5 yr	Food shortage, egg resorption		Ohgushi 1995
24.	*Conophthorus resinosae*	Coleoptera: Scolytidae	*Pinus resinosa*	U.S.A.	11 yr	Weather, food quantity		Mattson 1980
25.	*Dendroctonus ponderosae*	Coleoptera: Scolytidae	*Pinus* spp.	U.S.A.	14 yr	Forest health, tree stress		Berryman 1999

Table 10.1. (cont.)

Insect species	Taxonomic affinity	Host plant(s)	Locality	Study length	Major factor	$r^2 \times 100^a$	Reference
26. *Euura lasiolepis*	Hymenoptera: Tenthredinidae	*Salix lasiolepis*	U.S.A.	16 yr	Precipitation, resource quality	69%	Price *et al.* 1998a
27. *Aneugmenus* spp.	Hymenoptera: Tenthredinidae	*Pteridium aquilinum*	U.K.	19 yr	Disease? Plant quality?		Lawton 2000
28. *Strongylogaster lineata*	Hymenoptera: Tenthredinidae	*Pteridium aquilinum*	U.K.	11 yr	Plant quality?		Lawton 2000
29. *Diprion pini*	Hymenoptera: Diprionidae	*Pinus sylvestris*	Germany	188 yr	Plant stress? Pure stands, dry poor soils?		Klimetzek 1990
		Pinus sp.	France	50 yr	Parasitoids, diapause, starvation		Geri 1988
30. *Diprion hercyniae*	Hymenoptera: Diprionidae	*Picea glauca, P. mariana*	Canada	23 yr	Parasitoids, disease	60%	Neilsen and Morris 1964
31. *Choristoneura fumiferana*	Lepidoptera: Tortricidae	*Abies balsamea*	Canada	15 yr	Temperature, parasitoids, predators	79%	Morris 1963b
				36 yr	Climate, enemies, food web dynamics		Royama 1984, 1992
				Review	Landscape conditions		Mattson *et al.* 1988
32. *Tortrix viridana*	Lepidoptera: Tortricidae	*Quercus robur*	U.K.	16 yr	Resource limitation, variation	35%	Hunter *et al.* 1997

					Factor	%	Reference
33. Zeiraphera diniana	Lepidoptera: Tortricidae	Larix decidua	Switzerland	28 yr	Induced plant resistance		Baltensweiler and Fischlin 1988
34. Acleris variana	Lepidoptera: Tortricidae	Picea, Abies	Canada	12 yr	Heat available, parasitoids	92%	Miller 1966
35. Epiphyas postvittana	Lepidoptera: Tortricidae	Pyrus communis	Australia	5 yr, 16 gen	Egg and 1st instar predation	28–67%	Danthanarayana 1983
36. Epinotia tedella	Lepidoptera: Tortricidae	Picea	Denmark	19 yr	Parasitoids	73%	Munster-Swendsen 1985
37. Coleophora laricella	Lepidoptera: Coleophoridae	Larix decidua	Germany	9 yr	Reduced fecundity, larval mortality	79%, 52%	Jagsch 1973
38. Phyllonorycter tremuloidiella	Lepidoptera: Gracillariidae	Populus tremuloides	U.S.A.	9 yr	Host/insect phenological synchrony		Auerbach et al. 1995
39. Cameraria hamadryadella	Lepidoptera: Gracillariidae	Quercus alba, etc.	U.S.A.	13 yr	Long season for 2 gen/yr		Auerbach et al. 1995
40. Stilbosus quadricustatella	Lepidoptera: Cosmopterygidae	Quercus geminata	U.S.A.	6 yr	Host plant phenology		Auerbach et al. 1995
41. Operophtera brumata	Lepidoptera: Geometridae	Quercus robur	U.K.	16 yr	Mortality of young larvae	66%	Varley et al. 1973
42. Bupalus piniarius	Lepidoptera: Geometridae	Pinus		16 yr	Pupal predation	34%	Hunter et al. 1997
			Germany	188 yr	Pure stands, poor, dry soils?		Klimetzek 1990
			Germany	60 yr	Unknown		Schwerdtfeger 1935, 1941
			Netherlands	15 yr	First instar mortality		Klomp 1966
			U.K.	32 yr	Pupal weight, parasitoids		Barbour 1988

Table 10.1. (cont.)

Insect species	Taxonomic affinity	Host plant(s)	Locality	Study length	Major factor	$r^2 \times 100^a$	Reference
43. *Epirrita autumnata*	Lepidoptera: Geometridae	*Betula pubescens*	Finland	115 yr	Sunspot activity – weather in north, predation in south		Tenow 1972, Ruohomäki et al. 2000
44. *Malacosoma neustria*	Lepidoptera: Lasiocampidae	Hardwoods	Japan	8 yr	Reduced fecundity, pupal predation	69%, 36%	Shiga 1979
45. *Malacosoma californicum*	Lepidoptera: Lasiocampidae	*Alnus rubra*, etc	Canada	24 yr	Weather, viral disease		Myers 2000
46. *Malacosoma disstria*	Lepidoptera: Lasiocampidae	*Populus tremuloides*	U.S.A.	55 yr	Weather, parasitoids		Hodson 1941
47. *Dendrolimus pini*	Lepidoptera: Lasiocampidae	*Pinus*	Germany	60 yr	Unknown		Schwerdtfeger 1935, 1941
48. *Lymantria dispar*	Lepidoptera: Lymantriidae	Deciduous trees	U.S.A.	73 yr	Weather, oak mast, mammals, predation		Liebhold et al. 2000
49. *Lymantria monacha*	Lepidoptera: Lymantriidae	*Pinus*	Germany	188 yr	Pure stands, poor dry soils?		Klimetzek 1990
		Pinus, Picea	Denmark, Germany	90 yr	Dry, poor soils and warm dry summers		Bejer 1988
50. *Orgyia pseudotsugata*	Lepidoptera: Lymantriidae	*Pseudotsuga menziesii*	U.S.A.	70 yr	Natural enemies, warm, dry zones		Mason and Wickman 1988

Species	Order: Family	Host plant	Location	Duration	Mortality factor	Percentage	Reference
51. *Hyphantria cunea*	Lepidoptera: Arctiidae	Deciduous trees	Japan	6 yr	Larval mortality	66–73%	Itô and Miyashita 1968
		Deciduous trees	Canada	11 yr	Heat available (°D)		Morris 1969
52. *Tyria jacobaeae*	Lepidoptera: Arctiidae	*Senecio jacobaea*	U.K.	9 yr	Starvation/dispersal, pupal predation	44–64%, 81%	Dempster 1982
			U.K.	17 yr	Rainfall, food available	42%?	van der Meijden et al. 1998
			Netherlands	23 yr	Rainfall, food availablle	53%	van der Meijden et al. 1988
53. *Syntypistis punctatella*	Lepidoptera: Notodontidae	*Fagus* spp.	Japan	9 yr	Disease, site conditions		Kamata 2000
54. *Panolis flammea*	Lepidoptera: Noctuidae	*Pinus sylvestris*	Germany	188 yr	Pure stands, poor, dry soils?		Klimetzek 1990
		Pinus contorta		60 yr	Unknown		Schwerdtfeger 1935, 1941
			U.K.	17 yr	Parasitoids, disease?		Watt and Hicks 2000
55. *Hyloicus pinastri*	Lepidoptera: Sphingidae	*Pinus*	Germany	50 yr	Unknown		Schwerdtfeger 1935, 1941
56. *Andraca bipunctata*	Lepidoptera: Bombycidae	*Thea sinensis*	India	3 yr, 10 gen	Reduced fecundity	88%	Banerjee 1979

Table 10.1. (cont.)

Insect species	Taxonomic affinity	Host plant(s)	Locality	Study length	Major factor	$r^2 \times 100^a$	Reference
57. *Papilio xuthus*	Lepidoptera: Papilionidae	Citrus trees	Japan	4 yr,16 gen	Reduced fecundity	45%	Hirose et al. 1980
58. *Ladoga camilla*	Lepidoptera: Nymphalidae	*Lonicera periclymenum*	U.K.	7 yr	Predation on larvae and pupae	49%	Pollard 1979
59. *Anthocaris cardaminis*	Lepidoptera: Pieridae	*Cardamine pratensis*	U.K.	18 yr	Food quantity	52–85%	Dempster 1997
60. *Colias alexandra*	Lepidoptera: Pieridae	*Lathyrus leucanthus*	U.S.	5 yr	Reduced fecundity	96%	Hayes 1981
61. *Maculinea arion*	Lepidoptera: Lycaenidae	*Thymus, Origanum*	U.K.	5 yr	Larval mortality	29–42%	Thomas 1974
62. *Thecla betulae*	Lepidoptera: Lycaenidae	*Prunus spinosa*	U.K.	6 yr	Reduced fecundity	77%	Thomas 1974

[a] When available, the percentage of variance in population density accounted for by the independent variable(s) is provided.

3. Only a small proportion of studies are on sparsely distributed species, in addition to those mentioned in (2) above: spp. 16, 21, 23, 26, 61, and probably the other butterfly species, 57–60 and 62. The majority of the remaining species studied are pests, and most of them are on trees with moth species as the best-represented group (spp. 31–56), 26 species in all. The reason for these emphases are pragmatic, of course, but the need to understand individual pest species has not been augmented by any kind of conceptual basis that requires broad comparative studies, say of common and rare species in the same habitat. In some cases species have been studied in outbreak and non-outbreak areas (e.g. sp. 33, 43) but this has been rare relative to the evident opportunities and benefits.

4. Resource limitation in terms of food quantity and or quality appears in various guises: intraspecific competition in tree-dwelling aphids (spp. 2–6, 8); quantitative limitation of food in seed and cone feeders (spp. 16, 18, and 24); dispersal from crowded sites (spp. 52, 23, the aphids, and Dempster (1983) mentions many cases in the Lepidoptera, appearing in Table 10.1 as reduced fecundity); reduced fecundity (spp. 37, 44, 56, 57, 60, and 62); and phenological overlap of resource and herbivore (spp. 38–40). Not surprisingly food supply is crucial to many species, and probably all, but it often does not enter the catalogs of death generated in life tables. For example, Morris's (1963b) original model for spruce budworm (sp. 31) included predation, parasitism, and temperature as key ingredients, based on plot studies and life tables. Royama (1984, 1992) saw a much more complex picture with a landscape perspective, and Mattson *et al.* (1988) emphasized landscape conditions with an emphasis on suitable food trees in dense stands in favorable abiotic conditions. Such a conclusion was very evident to Morris and associates, illustrating a serious concern that idiosyncratic studies can reach very different conclusions depending on the methods employed. When mortality is measured in detail, and natality is largely ignored, the conclusions are circumscribed and inevitable (Price *et al.* 1990). And since food is always present or the insects would be unavailable for study, plant resources have been taken for granted and often ignored. I am confident that for every herbivore studied adequately, and with our current understanding of food quality variation, all herbivore species could be shown to be influenced profoundly by food supply.

5. The debate on bottom-up versus top-down effects on population dynamics was not developed as an either/or proposition, but rather

to emphasize the understudied importance of food supply (Hunter and Price 1992a, b). Dempster (1983, p. 478) had noted much earlier that "at present too few ecologists have quantified the carrying capacity of habitats occupied by the species they study..." and this no doubt remains true for many studies in Table 10.1. Comparing conclusions reached and tabulated here, and using the criteria for bottom-up, top-down, and lateral effects proposed by Harrison and Cappuccino (1995), with one exception, the following summarizes the results: 68 percent (42 studies) show bottom-up effects, and 40 percent (25 studies) show top-down effects. These results are in concordance with those for herbivorous insects derived from density-perturbation experiments summarized by Harrison and Cappuccino (1995): 67 percent of studies (total 9) showed bottom-up effects and 33 percent showed top-down effects (6 studies). Such similarities between mostly observational and experimental studies may be fortuitous, but both sets of data trend strongly toward a majority of bottom-up effects in insect herbivore population dynamics. Added to this are the effects of weather factors working through resource supply in quantity and or quality, adding another 8 percent (5 species) to the bottom-up effects (spp. 16, 24, 26, 39, and 48). The dubious inclusion of species 48 here must be noted, where weather affects oak mast, influencing small mammal numbers that act as predators on gypsy moth. But in many cases both bottom-up and top-down influences are important: 21 percent of species (13 studies). Direct effects of weather were noted for 15 percent of species (9 studies) and 8 percent (5 studies) were not understood at all.

The one instance in which I deviated from Harrison and Cappuccino's (1995) criteria for classifying the direction of influences concerns lateral effects in which they included dispersal. In the studies used in Table 10.1 dispersal is clearly related to crowding resulting from shortage of food and habitat, so I have included dispersal and its resultant loss of fecundity as a bottom-up effect.

As in (4) above, on food constituting probably ubiquitous importance, weather and climate are no doubt essential ingredients of a full understanding of population dynamics, as Andrewartha and Birch (1954) emphasized long ago. Both influence host plant and insect distributions, forcing their consideration in a fully geographic analysis of distribution, abundance, and population dynamics. Species studied near the edge of their geographic or altitudinal ranges show clear effects of weather on their dynamics (e.g. spp. 14, 26, and 51). As Walker and Jones (2001) note, the strength of effects up and down terrestrial trophic systems requires objective, pluralistic evaluation in the future development of synthesis.

6. Many formal sampling studies have missed some of the most interesting aspects of dynamics according to current points of view. For example, reasons for periodicity of fluctuations in density and spatial synchrony over large areas are now under close inspection, but conclusions are elusive (Myers 1988; Liebhold and Kamata 2000). And widening investigations well beyond the focal food web may result in fascinating new emphases, such as the weather, oak mast, small mammal and gypsy moth connections (Elkinton *et al.* 1996; Jones *et al.* 1998; Liebhold *et al.* 2000). Exogenous periodic weather factors may also be more critical than endogenous population interactions among herbivores and natural enemies (e.g. Williams and Liebhold 1995, 2000; Hunter and Price 1998). Long-term effects of woody plant age are clearly important in physiological stress-related susceptibility of host plant populations which are missed in many plot-sampling studies but noted in more geographic views of population dynamics (spp. 25, 29, 42, 49, and 54) where ramets age rapidly in clonal plants (sp. 26) and where annual rings in trees bear the marks of herbivore attack (sp. 19). In the last case, the unique study by Ylioja *et al.* (1999) recorded attack through cambium each year in the life of the trees, recorded by browning of the xylem by the insect damage, such that at the time of felling a record of attack each year for each tree was available. This is the only case where year-by-year population densities of an insect herbivore were recorded posthumously on each tree sampled, showing a clear tree-age-related trend in insect abundance per tree. The prospect is that many insect herbivores are influenced by plant age (e.g. Price *et al.* 1990; Kearsley and Whitham 1989).

7. Many more points could be discussed based on Table 10.1, but the overriding impression received is that an emphasis on ecological factors in population dynamics yields the full panoply of possible influences. Almost anything ecological can impinge on a species' dynamics, leaving predictability seriously challenged, discovery of pattern narrowly confined, and dooming idiosyncratic approaches to go their own individualistic ways, reaching their own conclusions. The potential of science remains unfulfilled! Does Table 10.1 represent "a pile of sundry facts – some of them interesting or curious but making no meaningful picture as a whole" (Dobzhansky 1973, p. 129)?

So this is the legacy of 200 years of studying insect populations: a cornucopia of idiosyncratic studies with uncertain understanding, and no theory. The longest-studied aspect of ecology has failed to develop as science should: toward a sound theoretical foundation, accounting mechanistically for the way in which insect populations behave in space and in time. Can a purely ecological approach achieve the necessary explanations of broad patterns in nature, or must a Darwinian view be conscripted, espoused, and generally adopted? Orians (1962, p. 262) stated clearly that "evolution would seem to be the only real theory of ecology today." He went on to say that "Even if one strongly believes in the action of natural selection it is exceedingly difficult, as Darwin pointed out, to keep it always firmly in mind. Neglect of natural selection in ecological thinking is, therefore, understandable though regrettable. However, its deliberate exclusion in these years following the Darwin centennial would seem to be exceedingly unwise." Orians was writing at a time of great activity in insect population research with life table development and analysis and single factor or key factor analysis well in place. I do not believe that there was a deliberate exclusion of evolutionary thinking, but rather a conviction that populations were influenced purely by ecological forces, not by evolutionary history or natural selection. And the pragmatic training of agricultural and forest entomologists and ecologists probably failed to instill an enduring fascination with evolutionary biology. Hence, researchers in these fields were unlikely to be reading the likes of Mayr (1961) on "Cause and effect in biology," Orians (1962) on "Natural selection and ecological theory," and Dobzhansky (1973) in "Nothing in biology makes sense except in the light of evolution." Indeed, there was no clear conceptual avenue for uniting ecology and evolution in population dynamics studies.

The research agenda or Kuhnian paradigm (Kuhn 1962), which I have called the microecological idiosyncratic approach to population dynamics, which relied exclusively on ecological factors for explaining distribution, abundance, and population dynamics, was unified to some extent by the 1960s (cf. Clark et al. 1967; Southwood 1968; den Boer and Gradwell 1970; Varley et al. 1973). Emphasis was on particular species in their ecological environment. Very little of a broad comparative nature in population dynamics emerged, very little detection of broad patterns was undertaken. As a result, an empirically based theory on broad dynamical patterns in nature did not emerge. A mechanistic understanding and synthesis of population dynamics was impossible while employing the Microecological Idiosyncratic Paradigm.

A CHALLENGE TO THE ECOLOGICAL VIEW

Does the Macroevolutionary Nomothetic Paradigm I have developed in this book offer more than the Microecological Idiosyncratic Paradigm and does it provide the basis for empirically founded pluralistic theory on population dynamics? Does the Macroevolutionary Nomothetic Paradigm even qualify in the Kuhnian sense of a way of viewing the world, with a research agenda adopted by a major segment of a field of endeavor? And as Krebs (1995, p. 1) asks in general: "Which paradigm is more useful in making testable predictions and solving the key problems of the day?" I will address these questions after making a clear distinction between much of what is called ecological theory today and the Darwinian approach to theory which resulted in the theory of evolution, briefly discussed in Chapter 1.

Much ecological theory is actually hypothetical in nature, and often based only on the sparsest kinds of empirical observations, and certainly not corroborated by extensive testing in the field. For example, many theories have been proposed about aspects of latitudinal gradients in species diversity, such as species being more specialized in the tropics than in temperate regions, there being more competition in the tropics, fewer vacant niches, more niche compression, and more biotic interaction. In every case empirical evidence is not consistent with theory or is inadequate for the largest taxa, the herbivorous insects, about which much is known (Price 1991a). Such theory belongs in the ranks of ecological hypothesis, which remains as inadequately tested and without a large body of empirical data consistent with the hypothesis. A demanding set of criteria for a theory was defined by Pickett et al. (1994, p. 100):

> For theories to be most useful, some large proportion of the potential richness of components must be present. The most useful theories will incorporate explicit assumptions, clear domain, clear concepts and definitions, a body of fact, confirmed generalization, laws, models, a framework with translation modes, and hypotheses. Not only must some large proportion of the components of theory be present, but the individual components must be well developed for a theory to be maximally useful. Development refers to exactness, empirical certainty, applicability to observation, and derivativeness of complex components. In addition, connections among components must be specified, since the components of theory gain meaning and utility only in the context of the whole theory.

Most ecological theories fail to meet these criteria, and mysteriously, the authors provide no examples of what they consider to be excellent theory in ecology. Does this mean there are none?

Claims have been made that a theory of population dynamics exists, especially relating to forest insect populations. Starting with five fundamental principles on population behavior – exponential growth, cooperation, competition, circular causality, and limiting factors – makes the basis for theory very general (e.g. Berryman 1991, 1997, 1999). But Berryman's insistence that dynamics are purely ecological in nature misses some important questions and leads in some dubious directions, in my view. The most fundamental question is on the mechanistic explanation for the existence of outbreak species and why they are clustered in certain taxonomic groups. If we have general theory on forest insect dynamics, what is the empirically based explanation for low populations in the majority of species? And, when natural selection is inevitably present in all populations, why do outbreak species in forests remain so vulnerable to parasitoids (cf. Berryman 1996). The theory, in fact, overemphasizes top-down, negative feedback effects, just as life table development has done, while alternatives are not considered, such as plant quality variation (e.g. Hunter and Price 1998) or weather (Williams and Liebhold 1995, 2000), or both. Many insect species have no parasitoids known and others are rarely parasitized, but populations remain limited. How does the theory cope with these? And shouldn't a general theory address a wider range of habitats and taxa than insects in forests? Berryman (1999) does indeed apply time-series analysis to a broad range of species from forest insects to sandhill cranes and rock lobsters, but his analytical techniques are only suggestive of what might be the mechanistic underpinnings of population change.

When empiricists study nature, empirical information is gathered and hypotheses are erected. But dogged determination to establish the hypothesis as well founded by increasing the breadth of the study, and collecting evidence on the general applicability of the idea, is usually wanting. As a result, in ecology, we have enormous numbers of hypotheses but usually we know little or nothing of how broadly applicable they are or whether they account mechanistically for broad patterns in nature. I would call this "the curse of ecology" – too many hypotheses, too little empirically and broadly based theory.

This "curse of ecology" stems largely from an absence, or severe paucity, of training in science in synthesis, pattern detection, and broadly comparative approaches, all essential components of a large-scale mechanistic understanding of nature. We are excellent reductionists and terrible synthesists by and large. Books on research methodology do not discuss synthesis and the development of theory, just the nuts and bolts of idiosyncratic research projects (e.g. Lumley and Benjamin 1994; Booth *et al.* 1995; Krebs 1999). Also, synthesis and empirically based theory are founded on broad knowledge and extensive data, emphases that run counter to the strongly reductionist approaches favored by lecturers, funding agencies, time frames of research projects, and the age groups active in research. Unremitting creativity required in modern research institutions rewards poorly the synthesizers, collectors of comparative data, and generators of empirically based theory. It takes too long and it is an easy target of criticism. Caution suggests the reductionist research program.

IS THERE A MACROEVOLUTIONARY PARADIGM?

Returning to the questions to be addressed on the Macroevolutionary Nomothetic Paradigm, it seems to be essential to establish that the term paradigm holds credibility in this context. Of course, I have merely copied the Hennigian phylogeneticists' view of the world for use in ecology and specifically for addressing the central issues of distribution, abundance, and population dynamics. This is not a novel move for it follows a tradition of phylogenetic analysis fostered by Hennig (1950, 1966) and promoted by many authors since, especially in systematics but also in ecology (e.g. Mitter *et al.* 1988, 1991; Harvey and Pagel 1991; Farrell and Mitter 1993; Wiegmann *et al.* 1993; Hunter 1995a, b; Farrell 1998, in press; Sequeira *et al.* 2000). I have not applied phylogenetic analysis rigorously because the data necessary are not available. First we would need phylogenies for every taxon discussed in this book, with the traits in question, such as oviposition behavior and plant module utilization, analyzed in terms of the relative plesiomorphic or apomorphic position of each. We would need much more detail on population dynamics so that traits relating to distribution, abundance, and dynamics could be ordered. The normal dichotomy of outbreak and nonoutbreak species is much too crude, for there are many kinds of each. A classification of dynamical types is needed and a very good start has been made by Berryman, who recognizes two major outbreak types, gradient and eruptive, and then subdivides these into four subtypes each (Berryman and Stark 1985; Berryman 1986, 1987). But equivalent approaches are needed for nonoutbreak species that are common to rare, with Rabinowitz's (1981) classification forming a basis perhaps (Fig. 7.10). And then we would need enough detail on each insect species to place it correctly in the classification. Perhaps this book will encourage development of more phylogenetic approaches to distribution, abundance, and dynamics in insects and other taxa. This would then form the basis for an examination of the extent to which evolutionary factors or ecological factors prevail. An excellent model in my view would be Hunt's (1999) approach to the evolution of sociality in wasps (Vespidae) following the emergence of necessary traits throughout the evolution of the Hymenoptera. An equivalent analysis mapping traits onto a phylogeny would be novel and heuristic if enough were known of relevant traits and dynamics for a phylogenetic group. That is a challenge for the future, although a start has been made in Chapter 5 on sawflies.

I am advocating a paradigm shift in population dynamics but I am not generating a new one. I suggest that we replace the old idiosyncratic paradigm with the newer, but well-established, phylogenetic paradigm. This shift will provide a basis for systematizing the wealth of idiosyncratic information on species generated over the last 200 years.

COMPONENTS OF THE MACROEVOLUTIONARY THEORY

Here I muster the evidence to argue that there are grounds for a macroevolutionary theory on the distribution, abundance, and population dynamics of organisms, with focus on insect herbivores.

Throughout the book definitions of terms have been stated clearly, and they are also listed in the glossary. The assumption, though not made explicit, is that the theory of evolution is accepted as fact by functioning biologists and that it accounts for both the relationships among organisms and their diversity. The concept developed we have called the Phylogenetic Constraints Hypothesis which I think is clearly explained. The domain of the hypothesis is principally the distribution, abundance, and population dynamics of insect herbivores but the hypothesis can be extended to any biological taxon on which sufficient data exist. Certain criteria developed by Pickett et al. (1994) in their definition of useful theory have been met.

A body of fact is another criterion of "useful" theory, and it is well developed in this book. Much detailed information is provided in Chapters 3 and 4 on the species that led us to the Phylogenetic Constraints Hypothesis, and a lot more is in the cited literature. Understanding of the distribution, abundance, and dynamics of *Euura lasiolepis* is based on long-term field observations and extensive experimental tests on behavior, willow growth and modular display, ovipositional preference–larval performance relationships, and the role of natural enemies. Bottom-up, lateral, and top-down influences on this herbivore have been evaluated thoroughly using both proximate and ultimate explanation. Testing the hypothesis on a broader scale, we extended studies in a comparative way to include 36 species of tenthredinid sawfly (plus two species in other families; Table 5.2), with 13 species demonstrated to respond positively in ovipositional preference and larval performance to vigorous willow modules (Table 5.3). Studies in the Untied States, Finland, and Japan spanned the Holarctic geographic range of the gall-inducing sawflies. We extended studies beyond the sawflies to other galling taxa, to include representative species from eight insect families in four insect orders, for a total of 28 species including those studied by independent investigators (Table 6.1). Using the literature, including many of our own studies, we extended the comparisons to nongalling species, listing 56 cases (Table 6.2). In Tables 6.1 and 6.2 we use examples from temperate and tropical latitudes showing that patterns of plant utilization consistent with the plant vigor hypothesis and the Phylogenetic Constraints Hypothesis are not limited geographically nor are they taxonomically restricted. We show repeatedly convergence of emergent properties in divergent taxa with phylogenetic constraints in common. These results on species with relatively patchy distributions and low abundance are contrasted with well-studied outbreak or eruptive species with very different phylogenetic constraints. Thirty-three species of forest lepidopterans show traits consistent with the phylogenetic constraints hypothesis applied to outbreak species (Table 7.1), as well as other taxa discussed in Chapter 7: mites, stick insects, grasshoppers, aphids, scales, spittlebugs, leaf and bark beetles, and other kinds of lepidopterans. A logical case is developed on the divergence of emergent properties in the common and pine sawflies (Chapter 8), which show a convergence in frequency of outbreak species when tenthredinids and diprionids attack conifers (Table 8.2). The body of facts consistent with the pluralistic theory is extensive. The facts confirm that generalization is broadly possible and that it is broad in relation to taxonomic groups and biogeography.

I could continue to address the criteria for theory set by Pickett *et al.* (1994) in relation to the body of information provided in this book, but I think it should be clear that many criteria for "useful" theory are met concerning the Phylogenetic Constraints Hypothesis.

One criterion I would add to the Pickett criteria is predictive ability of a theory. Certainly, the Phylogenetic Constraints Hypothesis has much predictive power. Once the flow of effects from phylogenetic constraint to adaptive syndrome to emergent properties is known for one species we can predict that the potential exists for related species to illustrate similar kinds of ecology in distribution, abundance, and population dynamics. We can predict under which circumstances insect species or guilds or genera will respond to agricultural crops, in the biological control of weeds, or in their suceptibility to the biological control of insect herbivores. We can predict that similar phylogenetic constraints among divergent taxa will result in similar population phenomena. And based on knowledge of phylogenetic constraints we can predict how species and higher taxa will respond to plant module structure, and whether species are likely to be generalists or specialists in relation to the range of host plants utilized effectively.

IS THE THEORY FALSIFIABLE?

We considered ways in which the Phylogenetic Constraints Hypothesis could be falsified in Price and Carr (2000), listing the kinds of results as examples rather than an exhaustive itemization.

1. Inability to discover strong phylogenetic influences in the population dynamics of taxa not considered in this book.

2. Inability to show that phylogenetic constraints are important in understanding population dynamics.

3. Rejection of the adaptive syndrome as a logical mechanistic evolved set of traits causally linking phylogenetic constraints and emergent properties.

4. Failure to demonstrate convergence in emergent properties when similar constraints are identified in independent taxa under apparently equivalent selective environments.

5. Frequent rejection of the hypothesis that traits act as constraints.

6. Phylogenetic analyses of taxa that show traits in the putative adaptive syndrome to be more plesiomorphic than the purported constraining traits.

7. Alternative hypotheses that account for more of the variation in distribution, abundance, and population dynamics would outcompete the Phylogenetic Constraints Hypothesis. We examined a few alternatives in Price and Carr (2000) and found them unconvincing. Others may be more successful.

SYNTHESIS

Fully integrating population dynamics with evolutionary theory, such that a large component of ecology is embraced within the evolutionary synthesis, is overdue. The macroevolutionary approach provides a broad comparative basis for studies and a conceptual framework for documenting mechanistic linkages from evolved traits to ecological outcomes.

The macroevolutionary view also forces a synthesis of behavior with population dynamics, an integration that is not well developed (e.g. Anholt 1997). And a fully rounded understanding of a phylogenetic group's emergent properties would require both proximate and ultimate explanations for behavior, and life history traits, with physiological and genetic mechanisms incorporated into a mature phylogenetic constraints theory for any one taxon. An essential component of what Krebs (1995, p. 1) calls his "mechanistic paradigm" is extensive experimental research which has been underemployed or ignored in so many population dynamics studies.

For the first time, I think, a systematic approach to population dynamics is available, of potential application to any insect group under study and probably to any taxon outside the insects: other invertebrates, vertebrates, and plants.

An exciting possibility awaits development when a full phylogenetic hypothesis of a group is accompanied by extensive knowledge of distribution, abundance, and population dynamics of taxon members. Then salient traits relevant to emergent properties could be mapped onto the phylogeny. The evolutionary basis for variation in population dynamics in a taxon could then be mapped out in a phylogenetically rigorous manner. Ordination of evolved traits and ecology on phylogenetic hypotheses presents a challenge but the phylogenetic constraints hypothesis offers the opportunity. We started in Chapter 5 to offer an example of how a phylogeny of gall-inducing sawflies developed using different plant modules and how this has resulted in different population dynamics (Fig. 5.16). Methods are available for more rigorous tests of the theory, but the essential data must be gathered first (e.g. Felsenstein 1985; Maddison 1990; Harvey and Pagel 1991; Garland et al. 1992; McPeek 1995). Perhaps the simplest conceptual approach, adopted by Mitter et al. (1988) and Wiegmann et al. (1993), would be to identify sister taxa with contrasting characteristics and to quantify the number of species in each taxon. The technical details however are considerable as discussed in Mitter et al. (1988), who compared the extent of adaptive radiation of phytophagous insect taxa with their nonphytophagous sister group. In a similar vein, Wiegmann *et al.* (1993) compared the species number in parasitic and nonparasitic sister taxa. Farrell *et al.* (1991) asked if plant taxa with latex or resin canals diversified more than sister groups lacking such canals. This **"method of multiple sister-group comparisons"** (Wiegmann *et al.* 1993, p. 738) could be applied to sister groups of herbivorous insects with divergent adaptive syndromes with resultant latent or eruptive population dynamics. The details of such an analysis are still to be resolved.

As in other areas of investigation, such as in the evolution of eusociality in the Hymenoptera, we are faced with "a morass of morphological, behavioral, and ecological preconditions, the relative importance of which are hard to quantify" (Anderson 1984, p. 182). In ecology we have a large number of hypotheses relevant to distribution, abundance, and population dynamics, and a very large number of species to which they may apply. Sorting among them to find the most convincing hypothesis or group, even at the idiosyncratic, single-species level, has been challenging and frequently unsatisfactory. However, as Hunt (1999) illustrated for the evolution of vespid eusociality, ordination of traits on a phylogeny of the Hymenoptera provided a coherent methodology for rank-ordering traits basic to the eusocial condition. Relevant to our hypothesis on the sawflies, Hunt (1999, p. 230) noted that the starting groundplan for his hypothesis was the sawfly traits of mandibles and the lepismatid ovipositor. "This sclerotized, rigid structure is present in some of the earliest (Devonian) wingless insects, but among Holometabola it is retained only by Hymenoptera." This reinforced our argument that the sawfly ovipositor is a plesiomorphic trait and a phylogenetic constraint, and it illustrates that recognition of such groundplan traits can be basic to the integration of evolutionary and ecological data sets of various kinds.

A cautious approach to evaluating the Phylogenetic Constraints Hypothesis as established theory would be stepwise, from the narrow perspective to the broad. The most limited application of the hypothesis concerns the phylogeny of the gall-inducing sawflies, their abundance, distribution and dynamics, summarized in Fig. 5.16. The close mechanistic links are well established, in my view, from ovipositor to plant module utilization to specificity in host plant species, module specificity, phenology, behavior, and on to the emergent properties. Therefore, the hypothesis qualifies as an empirically based mechanistic explanation of a broad pattern in nature: a theory.

Moving beyond the gall-inducing sawflies to free-feeding sawflies, other gall-inducing taxa, and other groups, more data are needed to test the Phylogenetic Constraints Hypothesis adequately. Much evidence is consistent with the hypothesis but more examples covering the systematic range in a taxon are needed. Regrettably, for any one taxon this will probably necessitate dedicated research over many years by one investigative group using similar methodologies on the phylogeny, behavior, life history, and ecology involving at least several species per taxon. Relying on published papers by many different authors using different approaches inevitably yields contributions more to the idiosyncratic paradigm than the nomothetic paradigm.

However, there is no reason for despair. Linking the phylogenetic paradigm with population dynamics has much to offer, even in pairwise comparisons of species. And any level of synthesis between evolutionary theory and population dynamics is to be encouraged. The Phylogenetic Constraints Hypothesis offers a logical intercourse between these formerly separated fields, a union that will undoubtedly bear much fruit.

Glossary

adaptationist program the view that all organismal characters are optimally designed by natural selection, although some may be compromized by tradeoffs with other adaptations (cf. *Bauplan*, Phylogenetic constraint)

adaptive syndrome a coordinated set of adaptations that maximize the ecological potential of a species within the confines of a phylogenetic constraint

Bauplan the basic, general design of an organism and members of the clade to which it belongs; the ground plan and integrated whole of an organism which is based on its phylogenetic history and the constraints on development and design which limit further evolutionary potential

death table the author's preferred name for a life table when the emphasis is usually on the causes of death in a cohort of organisms

determinate growth in conifers hypothesis the argument that the rapid flush and determinate growth of conifers offers specialist insects on young modules a narrow window of time during which attack can be successful

density dependence a decrease in population growth rate of species A with increasing density of species A resulting from, for example, increasing percentage mortality caused by natural enemies

ecological morphology comparative morphology that emphasizes the links between organismal construction and the consequences for ecology and evolution

emergent properties within the context of this book, ecological results of evolved characters in the phylogenetic constraint and adaptive syndrome of a taxon, involving the ecological features of distribution, abundance and population dynamics

eruptive population dynamics characteristic of species that can increase dramatically in density through time so that populations become damaging to host plant populations, varying in density by three to five orders of magnitude, for example

facilitation a mechanism involving positive feedback in which the action of one individual improves resource quality for another individual in the same generation (as when gall induction improves plant vigor in distal internodes beneficial to other gall-inducing females)

genet an individual plant genotype represented by growth from a singleseed, which may include one plant or many resulting from vegetative multiplication

hypothesis an idea in need of further testing or a concept of how nature may be explained, without a strong body of empirical evidence consistent with the concept

idiographic an approach that regards each species and its ecology as individual and unique (contrasts with "nomothetic")

Idiosyncratic Descriptive Paradigm the commonly employed approach to population studies which treats each species in isolation, in an idiographic manner, or as a special case

latent population dynamics characteristic of species usually with steady densities through time, with low levels of damage to host plant populations, and generally incapable of increasing explosively in density; population densities at one site are likely to vary by one or two orders of magnitude through time

larval performance the manner in which larvae respond to a given set of conditions, evaluated as survival, weight gain or loss per unit time, final weight of larva, pupa or adult, or fecundity of females

life table a tabulation of the numbers in a cohort of organisms and their decline as mortality factors reduce survival through time, often including the identity of the factors responsible

macroevolution evolution above the species level, including the origin of new species, genera, and families and the evolution of clades and phylogenies

Macroevolutionary Nomothetic Paradigm the paradigm advocated in this book emphasizing a strongly comparative, mechanistic explanation of patterns in distribution, abundance, and population dynamics of species; contrasts with the Microecological Descriptive Paradigm

Microecological Descriptive Paradigm another name for the Idiosyncratic Descriptive Paradigm to contrast it directly with the Macroevolutionary Nomothetic Paradigm

microecology the study of one species at a time, generally using a strongly reductionist approach

macroecology the study of broad patterns in ecology

module an architectural component of a plant that is repeated many times as the plant grows (a stem, a leaf, bud, flower, etc.) resulting in the form of the plant and the nature of resources available to herbivores

nomothetic an approach emphasizing generalities, broad comparative views and synthesis (contrasts with "idiographic")

ovipositional preference any selectivity by an ovipositing female resulting in nonrandom distribution of eggs, often applied to selection of host plant species or modules within a host plant

Panglossian paradigm Gould and Lewontin's (1979) alternative term for the adaptationist program, an approach to science that assumes all organismal traits are adaptive and optimally designed; based on Voltaire's *Candide* in which the overoptimistic Dr. Pangloss's view was that all is for the best in this best of possible worlds

phylogeny the history of a taxon showing a common ancestor and the origin of new units (species, genera, families, etc.) in chronological order to produce a branching pattern; usually presented as a phylogenetic hypothesis

phylogenetic constraint a critical plesiomorphic character, or set of characters, common to a major taxon, that limits the major adaptive developments in a lineage and thus the ecological options for members of the taxon

Plant Apparency Hypothesis the idea that long-lived plants are apparent to herbivores so selection is for general but metabolically expensive defenses, such as digestibility reducers (e.g. tannins, resins, lignins) while ephemeral plants are hard to find and evolve with metabolically cheap defenses such as alkaloids

Plant Vigor Hypothesis the idea that vigorous host plants or plant modules are beneficial to insect herbivores in relation to their survival in particular, but rate of weight gain, ultimate size or weight, or fecundity, or favorable sex ratio may also be involved

pluralistic theory theory that recognizes the diversity of nature, with a variety of inputs (e.g. species and/or environments) which will result in a variety of outcomes (e.g. evolved results)

population regulation the maintenance near equilibrium densities in a population resulting from density-dependent processes

proovigenesis the condition in which a female insect emerges with a full complement of developed eggs, ready for oviposition

ramet one individual or one main clonal component derived from the vegetative reproduction of one parental genotype

resource regulation a mechanism of positive feedback in a population in which the action of one individual improves resource quality for a subsequent generation (e.g. by pruning stems)

synovigenesis gradual production of mature eggs by a female through her lifetime

theory the mechanistic explanation of broad patterns in nature, with the patterns observed empirically in nature, the explanations factual, and the weight of empirical evidence consistent with the explanation

References

Abrahamson, W. G. (ed.). 1989. *Plant–animal interactions*. McGraw-Hill, New York.

Abrahamson, W. G., and M. Gadgil. 1973. Growth form and reproductive effort in goldenrods (*Solidago*, Compositae). *Am. Nat.* 107: 651–661.

Allee, W. C., A. E. Emerson, O. Park, T. Park, and K. P. Schmidt. 1949. *Principles of animal ecology*. Saunders, Philadelphia.

van Alphen, J. J. M., and L. E. M. Vet. 1986. An evolutionary approach to host-finding and selection. pp. 23–61. In J. Waage and D. Greathead (eds.). *Insect parasitoids*. Academic Press, San Diego.

Åman, I. 1984. Oviposition and larval performance of *Rhabdophaga terminalis* on *Salix* spp. with special consideration of bud size of host plants. *Entomol. Exp. Appl.* 35: 129–136.

Anderson, M. 1984. The evolution of eusociality. *Annu. Rev. Ecol. Syst.* 15: 165–189.

Anderson, R. S., R. B. Davis, N. G. Miller, and R. Stuckenrath. 1986. History of late- and post-glacial vegetation and disturbance around Upper South Branch Pond, northern Maine. *Can. J. Bot.* 64: 1977–1986.

Andrewartha, H. G., and L. C. Birch. 1954. *The distribution and abundance of animals*. University of Chicago Press, Chicago.

Angevine, M. W., and B. F. Chabot. 1979. Seed germination syndromes in higher plants. pp. 188–206. In O. T. Solbrig, S. Jain, G. B. Johnson, and P. H. Raven (eds.). *Topics in plant population biology*. Columbia University Press, New York.

Anholt, B. R. 1997. How should we test for the role of behaviour in population dynamics? *Evol. Ecol.* 11: 633–640.

Anonymous. 1982. *The locust and grasshopper agricultural manual*. Center for Overseas Pest Research, London.

Arnett, R. H. 1993. *American insects: A handbook of the insects of America north of Mexico*. Sandhill Crane Press, Gainesville.

Auer, C. 1961. Ergebnisse zwölfjähriger quantitativer Untersuchungen der Populationsbewegung des Grauen Lärchenwichlers (*Zeiraphera griseana* Hb.) im Oberengadin, 1949–1958. *Mitt. schweiz. Anst. forstl. Vers. Wes.* 37: 175–263.

Auerbach, M. J., E. F. Connor, and S. Mopper. 1995. Minor miners and major miners: Population dynamics of leaf-mining insects. pp. 83–110. In N. Cappuccino and P. W. Price (eds.). *Population dynamics: New approaches and synthesis*. Academic Press, San Diego.

Baker, R. R. 1972. Territorial behaviour of the nymphalid butterflies, *Aglais urticae* (L.) and *Inachis io* (L.). *J. Anim. Ecol.* 41: 459–469.

Baker, R. R. 1983. Insect territoriality. *Annu. Rev. Entomol.* 28: 65–89.

Baker, W. L. 1972. *Eastern forest insects*. U.S. Dept. Agr. For. Serv. Misc. Pub. 1175. U.S. Government Printing Office, Washington, D.C.

Bakker, K. 1964. Backgrounds and controversies about population theories and their terminologies. *Z. angew. Entomol.* 53: 187–208.

Balda, R. P., A. C. Kamil, and P. A. Bednekoff. 1996. Predicting cognitive capacity from natural history: Examples from four species of corvids. *Current Ornithol.* 13: 33–66.

Baltensweiler, W. 1968. The cyclic population dynamics of the grey larch tortrix, *Zeiraphera griseana* Hübner (= *Semasia diniana* Guenée) (Lepidoptera: Tortricidae). pp. 88–97. In T. R. E. Southwood (ed.). *Insect abundance*. Symp. R. Entomol. Soc. London 4. Royal Entomological Society, London.

Baltensweiler, W., and A. Fischlin. 1988. The larch budmoth in the Alps. pp. 331–351. In A. A. Berryman (ed.). *Dynamics of forest insect populations: Patterns, causes, implications*. Plenum Press, New York.

Banerjee, B. 1979. A key factor analysis of population fluctuations in *Andraca bipunctata* Walker (Lepidoptera: Bombycidae). *Bull. Entomol. Res.* 69: 195–201.

Barbosa, P., and J. C. Schultz (eds.). 1987. *Insect outbreaks*. Academic Press, San Diego.

Barbosa, P., V. Krischik, and D. Lance. 1989. Life history traits of forest-inhabiting flightless Lepidoptera. *Am. Midl. Nat.* 122: 262–274.

Barbour, D. A. 1988. The pine looper in Britain and Europe. pp. 291–308. In A. A. Berryman (ed.). *Dynamics of forest insects populations: Patterns, causes, implications*. Plenum Press, New York.

Begon, M., J. L. Harper, and C. R. Townsend. 1996. *Ecology: Individuals, populations and communities*. Blackwell Science, Oxford, U.K.

Bejer, B. 1988. The nun moth in European spruce forests. pp. 211–231. In A. A. Berryman (ed.). *Dynamics of forest insect populations: Patterns, causes, implications*. Plenum Press, New York.

Benson, R. B. 1963. Wear and damage of sawfly saws (Hymenoptera: Tenthredinidae). *Natulae Entomol.* 43: 137–138.

Bernays, E. A. 1971a. The vermiform larva of *Schistocerca gregaria* (Forskal): Form and activity (Insecta, Orthoptera). *Z. Morphol. Tiere* 70: 183–200.

Bernays, E. A. 1971b. Hatching in *Schistocerca gregaria* (Forskal) (Orthoptera, Acrididae). *Acrida* 1: 44–60.

Bernays, E. A. 1972a. The muscles of newly hatched *Schistocerca gregaria* larvae and their possible functions in hatching, digging and ecdysial movements (Insecta: Acrididae). *J. Zool. Lond.* 166: 144–158.

Bernays, E. A. 1972b. The intermediate moult (first ecdysis) of *Schistocerca gregaria* (Forskal). *Z. Morphol. Tiere* 71: 160–197.

Bernays, E. A. 1972c. Changes in the first instar cuticle of *Schistocerca gregaria* before and associated with hatching. *J. Insect Physiol.* 18: 897–912.

Berryman, A. A. 1982. Population dynamics of bark beetles. pp. 264–314. In J. B. Mitton and K. B. Sturgeon (eds.). *Bark beetles in North American conifers: A system for the study of evolutionary biology*. University of Texas Press, Austin.

Berryman, A. A. 1986. *Forest insects: Principles and practices of population management*. Plenum Press, New York.

Berryman, A. A. 1987. The theory and classification of outbreaks. pp. 3–30. In P. Barbosa and J. C. Schultz (eds.). *Insect outbreaks*. Academic Press, San Diego.

Berryman, A. A. (ed.). 1988. *Dynamics of forest insect populations: Patterns, causes, implications*. Plenum Press, New York.

Berryman, A. A. 1991. Population theory: An essential ingredient in pest prediction, management, and policy-making. *Am. Entomol.* 37: 138–142.

Berryman, A. A. 1996. What causes population cycles of forest Lepidoptera? *Trends in Ecol. Evol.* 11: 28–32.

Berryman, A. A. 1997. On the principles of population dynamics and theoretical models. *Am. Entomol.* 43: 147–151.

Berryman, A. A. 1999. *Principles of population dynamics and their application.* Stanley Thorne, Cheltenham, U.K.

Berryman, A. A., and R. W. Stark. 1985. Assessing the risk of forest insect outbreaks. *Z. angew. Entomol.* 99: 199–208.

Betts, E. 1976. Forecasting infestations of tropical migrant pests: The desert locust and the African armyworm. pp. 113–134. In R. C. Rainey (ed.). *Insect flight.* Symp. R. Entomol. Soc. London 7. Wiley, New York.

Birch, L. C. 1958. The role of weather in determining the distribution and abundance of animals. *Cold Spring Harbor Symp. Quant. Biol.* 22: 203–218.

Blais, J. R. 1952. The relationship of the spruce budworm (*Choristoneura fumiferana* Clem.) to the flowering condition of balsam fir (*Abies balsamea* (L.) Mill.) *Can. J. Zool.* 30: 1–29.

Blais, J. R. 1958. Effects of defoliation of spruce budworm on radial growth at breast height of balsam fir and white spruce. *For. Chron.* 34: 39–47.

Blais, J. R. 1962. Collection and analysis of radial growth data from trees as evidence of past spruce budworm outbreaks. *For. Chron.* 38: 474–484.

Blais, J. R. 1965. Spruce budworm outbreaks in the past three centuries in the Laurentide Park, Quebec. *For. Sci.* 11: 130–138.

Blais, J. R., and G. H. Parks. 1964. Interaction of evening grosbeak (*Hesperiphona vespertina*) and spruce budworm (*Choristoneura fumiferana* (Clem.)) in a localized budworm outbreak treated with DDT in Quebec. *Can. J. Zool.* 42: 1017–1024.

Blais, J. R., and P. W. Price. 1965. Further evidence of a relationship between spruce budworm and evening grosbeak populations. *Can. Dept. For. Bi-monthly Prog. Rep.* 21(3): 1.

Bodenheimer, F. S. 1925a. On predicting the developmental cycles of insects. i. *Ceratitis capitata*, *Wied. Bull. Soc. Entomol. Egypte* 1924: 149–157.

Bodenheimer, F. S. 1925b. Ueber die Voraussage der Generationenzahl von Insekten. ii. Die Temperaturentwicklungskurve bei medizinisch wichtigen Insekten. *Zbl. Bakt.* i. 93: 474–480.

Bodenheimer, F. S. 1926. Ueber die Voraussage der Generationenzahl von Insekten. iii. Die Bedeutung der Klimas für die landwirtschaftliche Entomologie. *Z. angew. Entomol.* 12: 91–122.

Bodenheimer, F. S. 1927. Über die ökologischen Grenzen der Verbreitung von *Calandra oryzae*, L., und *Calandra granaria*, L. *Z. Wiss. Insekt Biol.* 22: 65–73.

Bodenheimer, F. S. 1930. Über die Grundlagen einer allgemeinen Epidemiologie der Insektenkalamitäten. *Z. angew. Entomol.* 16: 433–450.

Bonner, J. T. 1965. *Size and cycle: An essay on the structure of biology.* Princeton University Press, Princeton.

Booth, W. C., G. G. Colomb, and J. M. Williams. 1995. *The craft of research.* University of Chicago Press, Chicago.

Brokaw, N. V. L. 1985. Treefalls, regrowth, and community structure in tropical forests. pp. 53–69. In S. T. A. Pickett and P. S. White (eds.). *The ecology of natural disturbance and patch dynamics.* Academic Press, San Diego.

Brown, J. H. 1981. Two decades of homage to Santa Rosalia: Toward a general theory of diversity. *Am. Zool.* 21: 877–888.

Brown, J. H. 1984. On the relationship between abundance and distribution of species. *Am. Nat.* 124: 255–279.

Brown, J. H. 1995. *Macroecology*. University of Chicago Press, Chicago.

Brown, J. H., and M. V. Lomolino. 1998. *Biogeography*. 2nd edn. Sinauer, Sunderland.

Brown, J. H., and B. A. Maurer. 1987. Evolution of species assemblages: Effects of energetic constraints and species dynamics on the diversification of North American avifauna. *Am. Nat.* 130: 1–17.

Brown, J. H., and B. A. Maurer. 1989. Macroecology: The division of food and space among species on continents. *Science* 243: 1145–1150.

Bush, G. L. 1975a. Sympatric speciation in phytophagous parasitic insects. pp. 187–206. In P. W. Price (ed.). *Evolutionary strategies of parasitic insects and mites*. Plenum Press, New York.

Bush, G. L. 1975b. Modes of animal speciation. *Annu. Rev. Ecol. Syst.* 6: 339–364.

Bush, G. L., and J. J. Smith. 1998. The genetics and ecology of sympatric speciation: A case study. *Res. Popul. Ecol.* 40: 175–187.

Calder, W. A. 1984. *Size, function and life history*. Harvard University Press, Cambridge.

Caouette, M. R., and P. W. Price. 1989. Growth of Arizona rose and attack and establishment of gall wasps *Diplolepis fusiformans* and *D. spinosa* (Hymenoptera: Cynipidae). *Environ. Entomol.* 18: 822–828.

Cappuccino, N., and P. W. Price (eds.). 1995. *Population dynamics: New approaches and synthesis*. Academic Press, San Diego.

Carr, T. G. 1995. Oviposition preference–larval performance relationships in three free-feeding sawflies. Masters thesis, Northern Arizona University, Flagstaff.

Carr, T. G., H. Roininen, and P. W. Price. 1998. Oviposition preference and larval performance of *Nematus oligospilus* (Hymenoptera: Tenthredinidae) in relation to host plant vigor. *Environ. Entomol.* 27: 615–625.

Caswell, H. 1988. Theory and models in ecology: A different perspective. *Ecological Modelling* 43: 33–44.

Chapman, R. F. 1998. *The insects: Structure and function*. 4th edn. Cambridge University Press, Cambridge, U.K.

Chapman, R. F., and G. A. Sword. 1997. Polyphagy in the Acridomorpha. pp. 183–195. In S. K. Gangwere, M. C. Muralirangan, and M. Muralirangan (eds.). *The bionomics of grasshoppers, katydids and their kin*. CAB International, Wallingford, U.K.

Chitty, D. 1957. Self-regulation of numbers through changes in viability. *Cold Spring Harbor Symp. Quant. Biol.* 22: 277–280.

Chitty, D. 1960. Population processes in the vole and their relevance to general theory. *Can. J. Zool.* 38: 99–113.

Chitty, D. 1967. The natural selection of self-regulating behaviour in animal populations. *Proc. Ecol. Soc. Austral.* 2: 51–78.

Choe, J. C., and B. J. Crespi (eds.). 1997a. *The evolution of mating systems in insects and arachnids*. Cambridge University Press, Cambridge, U.K.

Choe, J. C., and B. J. Crespi (eds.). 1997b. *The evolution of social behavior in insects and arachnids*. Cambridge University Press, Cambridge, U.K.

Clancy, K. M., P. W. Price, and T. P. Craig. 1986. Life history and natural enemies of an undescribed sawfly near *Pontania pacifica* (Hymenoptera: Tenthredinidae) that forms leaf galls on arroyo willow, *Salix lasiolepis*. *Ann. Entomol. Soc. Am.* 79: 884–892.

Clancy, K. M., P. W. Price, and C. F. Sacchi. 1993. Is leaf size important for a leaf-galling sawfly (Hymenoptera: Tenthredinidae)? *Environ. Entomol.* 22: 116–126.

Clark, L. R., P. W. Geier, R. D. Hughes, and R. F. Morris. 1967. *The ecology of insect populations in theory and practice.* Methuen, London.

Cole, L. C. 1957. Sketches of general and comparative demography. *Cold Spring Harbor Symp. Quant. Biol.* 22: 1–15.

Coley, P. D. 1983. Herbivory and defensive characteristics of tree species in a lowland tropical forest. *Ecol. Monogr.* 53: 209–233.

Coley, P. D., and T. M. Aide. 1991. Comparison of herbivory and plant defenses in temperate and tropical broad-leaved forests. pp. 25–49. In P. W. Price, T. M. Lewinsohn, G. W. Fernandes, and W. W. Benson (eds.). *Plant–animal interactions: Evolutionary ecology in tropical and temperate regions.* Wiley, New York.

Coley, P. D., and T. A. Kursar. 1996. Anti-herbivore defenses of young tropical leaves: Physiological constraints and ecological trade-offs. pp. 305–336. In S. S. Mulkey, R. L. Chazdon, and A. P. Smith (eds.). *Tropical forest plant ecophysiology.* Chapman and Hall, New York.

Coley, P. D., J. P. Bryant, and F. S. Chapin. 1985. Resource availability and plant antiherbivore defense. *Science* 230: 895–899.

Colinvaux, P. 1993. *Ecology 2.* Wiley, New York.

Collins, J. P. 1986. Evolutionary ecology and the use of natural selection in ecological theory. *J. Hist. Biol.* 19: 257–288.

Conlong, D. E. 1990. A study of pest-parasitoid relationships in natural habitats: An aid towards the biological control of *Eldana saccharina* (Lepidoptera: Pyralidae) in sugar cane. *Proc. S. Afr. Sugar Tech. Assoc.* June: 111–115.

Cornell, H. V., and B. A. Hawkins. 1995. Survival patterns and mortality sources of herbivorous insects: Some demographic trends. *Am. Nat.* 145: 563–593.

Courtney, S. P., and T. T. Kibota. 1990. Mother doesn't know best: Selection of hosts by ovipositing insects. pp. 161–188. In E. A. Bernays (ed.). *Insect–plant interactions*, vol. 2. CRC Press, Boca Raton.

Craig, T. P. 1994. Effects of intraspecific plant variation on parasitoid communities. pp. 205–227. In B. A. Hawkins and W. Sheehan (eds.). *Parasitoid community ecology.* Oxford University Press, Oxford, U.K.

Craig, T. P., and T. Ohgushi. In press. Preference and performance are correlated in the spittlebug, *Aphrophora pectoralis* (Homoptera: Cercopoidea) on four species of willows. *Ecol. Entomol.*

Craig, T. P., P. W. Price, and J. K. Itami. 1986. Resource regulation by a stem-galling sawfly on the arroyo willow. *Ecology* 67: 419–425.

Craig, T. P., P. W. Price, K. M. Clancy, G. M. Waring, and C. F. Sacchi. 1988a. Forces preventing coevolution in the three trophic level system: Willow, a gall-forming herbivore, and parasitoid. pp. 57–80. In K. Spencer (ed.). *Chemical mediation of coevolution.* Academic Press, San Diego.

Craig, T. P., J. K. Itami, and P. W. Price. 1988b. Plant wound compounds from oviposition scars used in host discrimination by a stem-galling sawfly. *J. Insect Behav.* 1: 343–356.

Craig, T. P., J. K. Itami, and P. W. Price. 1989. A strong relationship between oviposition preference and larval performance in a shoot-galling sawfly. *Ecology* 70: 1691–1699.

Craig, T. P., J. K. Itami, and P. W. Price. 1990a. The window of vulnerability of a shoot-galling sawfly to attack by a parasitoid. *Ecology* 71: 1471–1482.

Craig, T. P., J. K. Itami, and P. W. Price. 1990b. Intraspecific competition and facilitation by a shoot-galling sawfly. *J. Anim. Ecol.* 59: 147–159.

Craig, T. P., P. W. Price, and J. K. Itami. 1992. Facultative sex ratio shifts by a herbivorous insect in response to variation in host plant quality. *Oecologia* 92: 153–161.

Crespi, B. J., D. A. Carmean, and T. W. Chapman. 1997. Ecology and evolution of galling thrips and their allies. *Annu. Rev. Entomol.* 42: 51–71.

Cromartie, W. J. 1975a. The effect of stand size and vegetational background on the colonization of cruciferous plants by herbivorous insects. *J. Appl. Ecol.* 12: 517–533.

Cromartie, W. J. 1975b. Influence of habitat on colonization of collard plants by *Pieris rapae*. *Environ. Entomol.* 4: 783–784.

Danell, K., and K. Huss-Danell. 1985. Feeding by insects and hares on birches earlier affected by moose browsing. *Oikos* 44: 75–81.

Danthanarayana, W. 1983. Population ecology of the light brown apple moth, *Epiphyas postvittana* (Walker) (Lepidoptera: Tortricidae). *J. Anim. Ecol.* 52: 1–33.

Darwin, C. 1859. *On the origin of species by means of natural selection, or the preservation of favoured races in the struggle for life.* John Murray, London.

Darwin, C. 1877. *The different forms of flowers on plants of the same species.* John Murray, London.

DeBach, P. (ed.). 1964. *Biological control of insect pests and weeds.* Reinhold, New York.

DeBach, P. 1974. *Biological control by natural enemies.* Cambridge University Press, London.

Deevey, E. S. 1947. Life tables for natural populations of animals. *Q. Rev. Biol.* 22: 283–314.

Dempster, J. P. 1963. The population dynamics of grasshoppers and locusts. *Biol. Rev. Cambridge Philos. Soc.* 38: 490–529.

Dempster, J. P. 1982. The ecology of the cinnabar moth, *Tyria jacobaeae* L. (Lepidoptera: Arctiidae). *Adv. Ecol. Res.* 12: 1–36.

Dempster, J. P. 1983. The natural control of populations of butterflies and moths. *Biol. Rev.* 58: 461–481.

Dempster, J. P. 1997. The role of larval food resources and adult movement in the population dynamics of the orange-tip butterfly (*Anthocaris cardamines*). *Oecologia* 111: 549–556.

Dempster, J. P., and I. F. G. McLean (eds.). 1998. *Insect populations: In theory and in practice.* Kluwer, Dordrecht, The Netherlands.

den Boer, P. J., and G. R. Gradwell (eds.). 1970. *Dynamics of populations.* Centre for Agricultural Publishing and Documentation, Wageningen, The Netherlands.

den Boer, P. J., and J. Reddingius. 1996. *Regulation and stabilization in population ecology.* Chapman and Hall, London.

Denno, R. F., and M. S. McClure (eds.). 1983. *Variable plants and herbivores in natural and managed systems.* Academic Press, San Diego.

Denno, R. F., C. Gratton, H. Döbel, and D. L. Finke. In press. Predation risk influences relative strength of top-down and bottom-up impacts in a guild of phytophagous insects.

Derrickson, E. M., and R. E. Ricklefs. 1988. Taxon-dependent diversification of life-history traits and the perception of phylogenetic constraints. *Funct. Ecol.* 2: 417–423.

Dial, K. P., and J. M. Marzluff. 1988. Are the smallest organisms the most diverse? *Ecology* 69: 1620–1624.

Dingle, H. 1996. *Migration: The biology of life on the move.* Oxford University Press, New York.

Dixon, A. F. G. 1973. *Biology of aphids.* Edward Arnold, London.

Dixon, A. F. G. 1990. Population dynamics and abundance of deciduous tree-dwelling aphids. pp. 11–23. In A. D. Watt, S. R. Leather, M. D. Hunter, and N. A. C. Kidd (eds.). *Population dynamics of forest insects.* Intercept, Andover, U.K.

Dixon, A. F. G. 1994. Individuals, populations and patterns. pp. 449–476. In S. R. Leather, A. D. Watt, N. J. Mills, and K. F. A. Walters (eds.). *Individuals, populations and patterns in ecology.* Intercept, Andover, U.K.

Dixon, A. F. G., P. Kindlmann, J. Leps, and J. Holman. 1987. Why there are so few species of aphids, especially in the tropics. *Am. Nat.* 129: 580–592.

Dobzhansky, T. 1973. Nothing in biology makes sense except in the light of evolution. *Am. Biol. Teacher* 35: 125–129.

Dodge, K. L., and P. W. Price. 1991a. Life history of the leaf beetle, *Disonycha pluriligata* (Coleoptera: Chrysomelidae), and host plant relationships with *Salix exigua* (Salicaceae). *Ann. Entomol. Soc. Am.* 84: 248–254.

Dodge, K. L., and P. W. Price. 1991b. Eruptive versus noneruptive species: A comparative study of host plant use by a sawfly, *Euura exiguae* (Hymenoptera: Tenthredinidae) and a leaf beetle, *Disonycha pluriligata* (Coleoptera: Chrysomelidae). *Environ. Entomol.* 20: 1129–1133.

Dodge, K. L., P. W. Price, J. Kettunen, and J. Tahvanainen. 1990. Preference and performance of the leaf beetle *Disonycha pluriligata* (Coleoptera: Chrysomelidae) in Arizona, and comparisons with beetles in Finland. *Environ. Entomol.* 19: 905–910.

Drake, V. A., and A. G. Gatehouse (eds.). 1995. *Insect migration: Tracking resources through space and time.* Cambridge University Press, Cambridge, U.K.

Dreger-Jauffret, F., and J. D. Shorthouse. 1992. Diversity of gall-inducing insects and their galls. pp. 8–33. In J. D. Shorthouse and O. Rohfritsch (eds.). *Biology of insect-induced galls.* Oxford University Press, New York.

Drolet, J., and J. N. McNeil. 1984. Performance of the alfalfa blotch leaf miner, *Agromyza frontella* (Rond.) (Diptera: Agromyzidae), on four alfalfa varieties. *Can. Entomol.* 116: 795–800.

Drooz, A. T. 1960. The larch sawfly: Its biology and control. *U.S. Dept. Agr. Tech. Bull.* 1212: 1–52.

Eaton, C. B. 1942. Biology of the weevil, *Cylindrocapturus eatoni* Buchanan, injurious to ponderosa and Jeffrey pine reproduction. *J. Econ. Entomol.* 35: 20–25.

Eckhardt, R. C. 1979. The adaptive syndromes of two guilds of insectivorous birds in the Colorado Rocky Mountains. *Ecol. Monogr.* 49: 129–149.

Elkinton, J. S., W. M. Healy, J. P. Buonaccorsi, G. H. Boettner, A. M. Hazzard, H. R. Smith, and A. M. Liebhold. 1996. Interactions among gypsy moths, white-footed mice, and acorns. *Ecology* 77: 2332–2342.

English-Loeb, G. M. 1989. Nonlinear responses of spider mites to drought-stressed host plants. *Ecol. Entomol.* 14: 45–55.

English-Loeb, G. M. 1990. Plant drought stress and outbreaks of spider mites: A field test. *Ecology* 71: 1401–1411.

Faegri, K., and L. van der Pijl. 1971. *The principles of pollination ecology.* Pergamon Press, Oxford, U.K.

Farrell, B. D. 1998. "Inordinate fondness" explained: Why are there so many beetles? *Science* 281: 555–559.

Farrell, B. D. In press. Evolutionary assembly of the milkweed fauna: Cytochrome oxidase 1 and the age of *Tetraopes* beetles. *Mol. Phylogenet. Evol.* 18.

Farrell, B. D., and C. Mitter. 1993. Phylogenetic determinants of insect/plant community diversity. pp. 253–266. In R. E. Ricklefs and D. Schluter (eds.). *Species diversity in ecological communities.* University of Chicago Press, Chicago.

Farrell, B., C. Mitter, and D. Dussourd. 1991. Macroevolution of plant defense: Do latex/resin secretory canals spur diversification? *Am. Nat.* 138: 881–900.

Fay, P. A., and T. G. Whitham. 1990. Within-plant distribution of a galling adelgid (Homoptera: Adelgidae): The consequences of conflicting survivorship, growth and reproduction. *Ecol. Entomol.* 15: 245–254.

Feeny, P. P. 1970. Seasonal changes in oak leaf tannins and nutrients as a cause of spring feeding by winter moth caterpillars. *Ecology* 51: 565–581.

Feeny, P. P. 1975. Biochemical coevolution between plants and their insect herbivores. pp. 3–19. In L. E. Gilbert and P. H. Raven (eds.). *Coevolution of animals and plants.* University of Texas Press, Austin.

Feeny, P. P. 1976. Plant apparency and chemical defense. pp. 1–40. In J. W. Wallace, and R. L. Munsell (eds.). *Biochemical interaction between plants and insects.* Plenum Press, New York.

Felsenstein, J. 1985. Phylogenies and the comparative method. *Am. Nat.* 125: 1–15.

Fernandes, G. W., and P. W. Price. 1988. Biogeographical gradients in galling species richness: Tests of hypotheses. *Oecologia* 76: 161–167.

Fernandes, G. W., and P. W. Price. 1991. Comparison of tropical and temperate galling species richness: The roles of environmental harshness and plant nutrient status. pp. 91–115. In P. W. Price, T. M. Lewinsohn, G. W. Fernandes, and W. W. Benson (eds.). *Plant–animal interactions: Evolutionary ecology in tropical and temperate regions.* Wiley, New York.

Ferrier, S. M. 1999. The significance of a rare bud-galling sawfly oviposition preference on willow. Masters thesis, Northern Arizona University, Flagstaff.

Finnegan, R. J. 1959. The pales weevil, *Hylobius pales* (Hbst.) in southern Ontario. *Can. Entomol.* 91: 664–670.

Finnegan, R. J. 1962. The pine root-collar weevil, *Hylobius radicis* Buch. in southern Ontario. *Can. Entomol.* 94: 11–17.

Flamm, R. O., R. N. Coulson, and T. L. Payne. 1988. The southern pine beetle. pp. 531–553. In A. A. Berryman (ed.). *Dynamics of forest insect populations: Patterns, causes, implications.* Plenum Press, New York.

Fondriest, S. M., and P. W. Price. 1996. Oviposition site resource quantity and larval establishment for *Orellia occidentalis* (Diptera: Tephritidae) on *Cirsium wheeleri. Environ. Entomol.* 25: 321–326.

Fowler, S. V. 1985. Differences in insect species richness and faunal composition of birch seedlings, saplings and trees: The importance of plant architecture. *Ecol. Entomol.* 10: 159–169.

Frankie, G. W., and D. L. Morgan. 1984. Role of host plant and parasites in regulating insect herbivore abundance, with emphasis on gall-inducing insects. pp. 101–140. In P. W. Price, C. N. Slobodchikoff, and W. S. Gaud (eds.). *A new ecology: Novel approaches to interactive systems.* Wiley, New York.

Fritsch, F. E., and E. Salisbury. 1938. *Plant form and function.* Bell, London.

Fritz, R. S., and P. W. Price. 1988. Genetic variation among plants and insect community structure: Willows and sawflies. *Ecology* 69: 845–856.

Fritz, R. S., and P. W. Price. 1990. A field test of interspecific competition on oviposition of gall-forming sawflies on willow. *Ecology* 71: 99–106.

Fritz, R. S., C. F. Sacchi, and P. W. Price. 1986. Competition versus host plant phenotype in species composition: Willow sawflies. *Ecology* 67: 1608–1618.

Fritz, R. S., B. A. Crabb, and C. G. Hochwender. 2000. Preference and performance of a gall-inducing sawfly: A test of the plant vigor hypothesis. *Oikos* 89: 555–563.

Furniss, R. L., and V. M. Carolin. 1977. *Western forest insects.* U.S. Dept. Agr. For. Serv. Misc. Pub. 1339. U.S. Government Printing Office, Washington, D.C.

Gadgil, M., and O. T. Solbrig. 1972. The concept of *r*- and *K*-selection: Evidence from wild flowers and some theoretical considerations. *Am. Nat.* 106: 13–31.

Gagné, R. J. 1989. *The plant-feeding gall midges of North America.* Cornell University Press, Ithaca.

Garland, T., P. H. Harvey, and A. R. Ives. 1992. Procedures for the analysis of comparative data using phylogenetically independent contrasts. *Syst. Biol.* 41: 18–32.

Gaston, K. J. 1988. Patterns in the local and regional dynamics of moth populations. *Oikos* 53: 49–57.

Gaston, K. J. 1994. *Rarity.* Chapman and Hall, London.

Gaston, K. J., and T. M. Blackburn. 1999. A critique for macroecology. *Oikos* 84: 353–368.

Gaston, K. J., and J. H. Lawton. 1988a. Patterns in the distribution and abundance of insect populations. *Nature* 331: 709–712.

Gaston, K. J., and J. H. Lawton. 1988b. Patterns in body size, population dynamics and regional distribution of bracken herbivores. *Am. Nat.* 132: 662–680.

Gaston, K. J., T. R. New, and M. J. Samways (eds.). 1993. *Perspectives on insect conservation.* Intercept, Andover, U.K.

Gauld, I., and B. Bolton (eds.). 1988. *The Hymenoptera.* British Museum (Natural History), London.

Geri, C. 1988. The pine sawfly in central France. pp. 377–405. In A. A. Berryman (ed.). *Dynamics of forest insect populations: Patterns, causes, implications.* Plenum Press, New York.

Gilbert, L. E. 1975. Ecological consequences of a coevolved mutualism between butterflies and plants. pp. 210–240. In L. E. Gilbert and P. H. Raven (eds.). *Coevolution of animals and plants.* University of Texas Press, Austin.

Gilbert, L. E. 1991. Biodiversity of a Central American *Heliconius* community: Pattern, process and problems. pp. 403–427. In P. W. Price, T. M. Lewinsohn, G. W. Fernandes, and W. W. Benson (eds.). *Plant–animal interactions: Evolutionary ecology in tropical and temperate regions.* Wiley, New York.

Gilpin, M., and I. Hanski (eds.). 1991. *Metapopulation dynamics: Empirical and theoretical investigations.* Academic Press, San Diego.

Godfray, H. C. J. 1994. *Parasitoids: Behavioral and evolutionary ecology.* Princeton University Press, Princeton.

Godfray, H. C. J., and C. B. Müller. 1998. Host–parasitoid dynamics. pp. 135–165. In J. P. Dempster and I. F. G. McLean (eds.). *Insect populations: In theory and in practice.* Kluwer, Dordrecht, The Netherlands.

Gould, S. J., and R. C. Lewontin. 1979. The spandrels of San Marco and the Panglossian paradigm: A critique of the adaptationist programme. *Proc. R. Soc. B* 205: 581–598.

Govindachari, T. R., and G. Suresh. 1997. Phytochemicals in locust and grasshopper management strategies. pp. 407–419. In S. K. Gangwere, M. C. Muralirangan, and M. Muralirangen (eds.). *The bionomics of grasshoppers, katydids and their kin.* CAB International, Wallingford, U.K.

Green, S. 2001. The caterpillar coin and a cautionary tale. *Antenna* 25: 157–159.

Grime, J. P. 1977. Evidence for the existence of three primary strategies in plants and its relevance to ecological and evolutionary theory. *Am. Nat.* 111: 1169–1194.

Grime, J. P. 1979. *Plant strategies and vegetation processes.* Wiley, Chichester, U.K.

Grissino-Mayer, H. D. 1996. A 2129-year reconstruction of precipitation for northwestern New Mexico, USA. pp. 191–204. In *Tree Rings, Environment, and Humanity: Proceedings of the International Conference,* Tucson, Arizona, 17–21 May 1994, Radiocarbon.

Grubb, P. J. 1977. The maintenance of species-richness in plant communities: The importance of the regeneration niche. *Biol. Rev.* 52: 107–145.

Haack, R. A., and W. J. Mattson. 1993. Life history patterns of North American tree-feeding sawflies. pp. 503–545. In M. Wagner and K. F. Raffa (eds.). *Sawfly life history adaptations to woody plants.* Academic Press, San Diego.

Hanski, I. A., and M. E. Gilpin (eds.). 1997. *Metapopulation biology: Ecology, genetics, and evolution.* Academic Press, San Diego.

Harcourt, D. G. 1969. The development and use of life tables in the study of natural insect populations. *Annu. Rev. Entomol.* 14: 175–196.

Harper, J. L. 1977. *Population biology of plants.* Academic Press, London.

Harper, J. L. 1982. After description. pp. 11–25. In E. I. Newman (ed.). *The plant community as a working mechanism.* Blackwell Scientific, Oxford, U.K.

Harper, J. L., and J. Ogden. 1970. The reproductive strategy of higher plants. I. The concept of strategy with special reference to *Senecio vulgaris* L. *J. Ecol.* 58: 681–698.

Harrison, S., and N. Cappuccino. 1995. Using density-manipulation experiments to study population regulation. pp. 131–147. In N. Cappuccino and P. W. Price (eds.). *Population dynamics: New approaches and synthesis.* Academic Press, San Diego.

Harvey, P. H., and M. D. Pagel. 1991. *The comparative method in evolutionary biology.* Oxford University Press, Oxford, U.K.

Hassell, M. P. 1970. Parasite behaviour as a factor contributing to the stability of insect host–parasite interactions. pp. 366–379. In P. J. den Boer and G. R. Gradwell (eds.). *Dynamics of populations.* Centre for Agricultural Publications and Documentation, Wageningen, The Netherlands.

Hassell, M. P. 2000. *The spatial and temporal dynamics of host–parasitoid interactions.* Oxford University Press, Oxford, U.K.

Hassell, M. P., and G. C. Varley. 1969. New inductive population model for insect parasites bearing on biological control. *Nature* 223: 1133–1137.

Hassell, M. P., M. J. Crawley, H. C. J. Godfray, and J. H. Lawton. 1998. Top-down versus bottom-up and the Ruritanian bean bug. *Proc. Natl Acad. Sci. U.S.A.* 95: 10661–10664.

Hawkins, B. A. 1988. Species diversity in the third and fourth trophic levels: Patterns and mechanisms. *J. Anim. Ecol.* 57: 137–162.

Hawkins, B. A. 1994. *Pattern and process in host–parasitoid interactions.* Cambridge University Press, Cambridge, U.K.

Hawkins, B. A., and J. H. Lawton. 1987. Species richness for parasitoids of British phytophagous insects. *Nature* 326: 788–790.

Hayes, J. L. 1981. The population ecology of a natural population of the pierid butterfly *Colias alexandra. Oecologia* 49: 188–200.

Hennig, W. 1950. *Grundzüge einer Theorie der phylogenetischen Systematik.* Deutscher Zentralverlag, Berlin.

Hennig, W. 1966. *Phylogenetic systematics.* University of Illinois Press, Urbana.

Herms, D. A., and W. J. Mattson. 1992. The dilemma of plants: To grow or defend? *Q. Rev. Biol.* 67: 301–352.

Herms, D. A., and W. J. Mattson. 1994. Plant growth and defense. *Trends Ecol. Evol.* 9: 488–489.

Hirose, Y., I. Suzuki, M. Takagi, K. Hiehata, M. Yamasaki, H. Kimoto, M. Yamanaka, M. Iga, and K. Yamaguchi. 1980. Population dynamics of the citrus swallowtail *Papilio xuthus* Linne (Lepidoptera: Papilionidae). Mechanisms stabilizing its numbers. *Res. Popul. Ecol.* 21: 260–285.

Hjältén, J., and P. W. Price. 1996. The effect of pruning on willow growth and sawfly population densities. *Oikos* 77: 549–555.

Hjältén, J., and P. W. Price. 1997. Can plants gain protection from herbivory by association with unpalatable neighbours? A field experiment in a willow-sawfly system. *Oikos* 78: 317–322.

Hodson, A. C. 1941. An ecological study of the forest tent caterpillar, *Malacosoma disstria* Hbn., in northern Minnesota. *Tech. Bull. Minn. Agric. Exp. Sta.* 148: 1–55.

Holdridge, L. R. 1947. Determination of world plant formations from simple climatic data. *Science* 105: 367–368.

Holdridge, L. R. 1967. *Life zone ecology*. 2nd edn. Tropical Research Center, San José, Costa Rica.

Holdridge, L. R., W. C. Grenke, W. H. Hatheway, T. Liang, and J. A. Tosi. 1971. *Forest environments in tropical life zones: A pilot study*. Pergamon Press, Oxford.

Houseweart, M. W., and H. M. Kulman. 1976. Fecundity and parthenogenesis of the yellowheaded spruce sawfly, *Pikonema alaskensis*. *Ann. Entomol. Soc. Am.* 69: 748–750.

Howard, L. O. 1897. A study in insect parasitism: A consideration of the parasites of the white-marked tussock moth, with an account of their habits and interrelations and with descriptions of new species. *Tech. Ser. U.S. Dept. Agr.* 5: 5–57.

Howard, L. O., and W. F. Fiske. 1911. The importation into the United States of the parasites of the gipsy moth and brown-tail moth. *Bull. U.S. Bur. Entomol.* 91: 1–344.

Howe, H. F., and L. C. Westley. 1986. Ecology of pollination and seed dispersal. pp. 185–215. In M. J. Crawley (ed.). *Plant ecology*. Blackwell Scientific Publications, Oxford, U.K.

Howe, H. F., and L. C. Westley. 1988. *Ecological relationships of plants and animals*. Oxford University Press, Oxford, U.K.

Hunt, J. H. 1999. Trait mapping and salience in the evolution of eusocial vespid wasps. *Evolution* 53: 225–237.

Hunter, A. F. 1991. Traits that distinguish outbreaking and non-outbreaking Macrolepidoptera feeding on northern hardwood trees. *Oikos* 60: 275–282.

Hunter, A. F. 1995a. The ecology and evolution of reduced wings in forest macrolepidoptera. *Evol. Ecol.* 9: 275–287.

Hunter, A. F. 1995b. Ecology, life history, and phylogeny of outbreak and nonoutbreak species. pp. 41–64. In N. Cappuccino and P. W. Price (eds.). *Population dynamics: New approaches and synthesis*. Academic Press, San Diego.

Hunter, M. D. 1990. Differential susceptibility to variable plant phenology and its role in competition between two insect herbivores on oak. *Ecol. Entomol.* 15: 401–408.

Hunter, M. D. 1992a. A variable insect–plant interaction: The relationship between tree budburst phenology and population levels of insect herbivores among trees. *Ecol. Entomol.* 17: 91–95.

Hunter, M. D. 1992b. Interactions within herbivore communities mediated by host plant: The keystone herbivore concept. pp. 287–325. In M. D. Hunter, T. Ohgushi, and P. W. Price (eds.). *Effects of resource distribution on animal–plant interactions*. Academic Press, San Diego.

Hunter, M. D., and P. W. Price. 1992a. Playing chutes and ladders: Heterogeneity and the relative roles of bottom-up and top-down forces in natural communities. *Ecology* 73: 724–732.

Hunter, M. D., and P. W. Price 1992b. Natural variability in plants and animals. pp. 1–12. In M. D. Hunter, T. Ohgushi, and P. W. Price. (eds.). *Effects of resource distribution on animal–plant interactions*. Academic Press, San Diego.

Hunter, M. D., and P. W. Price. 1998. Cycles in insect populations: Delayed density dependence or exogenous driving variables? *Ecol. Entomol.* 23: 216–222.

Hunter, M.D., T. Ohgushi, and P. W. Price (eds.). 1992. *Effects of resource distribution on animal-plant interactions.* Academic Press, San Diego.

Hunter, M. D., G. C. Varley, and G. R. Gradwell. 1997. Estimating the relative roles of top-down and bottom-up forces on insect herbivore populations: A classic study revisited. *Proc. Natl Acad. Sci. U.S.A.* 94: 9176–9181.

Hutchinson, G. E. 1957. Concluding remarks. *Cold Spring Harbor Symp. Quant. Biol.* 22: 415–427.

Hutchinson, G. E. 1965. *The ecological theater and the evolutionary play.* Yale University Press, New Haven.

Inbar, M., H. Doostdar, and R. T. Mayer. 2001. Suitability of stressed and vigorous plants to various insect herbivores. *Oikos* 94: 228–235.

Itô, Y., and K. Miyashita. 1968. Biology of *Hyphantria cunea* in Japan. V. Preliminary life tables and mortality data in urban areas. *Res. Popul. Ecol.* 10: 177–209.

Jackson, J. B. C., L. W. Buss, and R. E. Cook (eds.). 1985. *Population biology and evolution of clonal organisms.* Yale University Press, New Haven.

Jagsch, A. 1973. Populationsdynamik und Parasitenkomplex der Lärchenminiermotte, *Coleophora laricella* Hbn., in natürlichen Verbreit ungsgebiet der europäischen Lärche, *Larix decidua* Mill. *Z. angew. Entomol.* 73: 1–42.

Johnson, C. R., and M. C. Boerlijst. 2002. Selection at the level of the community: The importance of spatial structure. *Trends Ecol. Evol.* 17: 83–90.

Johnson, W. T., and H. H. Lyon. 1976. *Insects that feed on trees and shrubs: An illustrated practical guide.* Cornell University Press, Ithaca.

Jones, C. G., R. S. Ostfeld, M. P. Richard, E. M. Schauber, and J. O. Wolff. 1998. Chain reactions linking acorns to gypsy moth outbreaks and Lyme disease risk. *Science* 279: 1023–1026.

Jones, T. H., M. P. Hassell, and H. C. J. Godfray. 1997. Host–multiparasitoid interactions. pp. 257–275. In A. C. Gange and V. K. Brown (eds.). *Multitrophic interactions in terrestrial systems.* Blackwell Science, Oxford, U.K.

Kamata, N. 2000. Population dynamics of the beech caterpillar, *Syntypistis punctatella,* and biotic and abiotic factors. *Popul. Ecol.* 42: 267–278.

Karban, R. 1987. Herbivory dependent on plant age: A hypothesis based on acquired resistance. *Oikos* 48: 336–337.

Karban, R. 1990. Herbivore outbreaks on only young trees: Testing hypotheses about aging and induced resistance. *Oikos* 59: 27–32.

Karban, R., and I. T. Baldwin. 1997. *Induced responses to herbivory.* University of Chicago Press, Chicago.

Kearsley, M. C., and T. G. Whitham. 1989. Developmental changes in resistance to herbivory: Implications for individuals and populations. *Ecology* 70: 422–434.

Key, K. H. L. 1991. Phasmatodea (stick-insects). pp. 394–404. In I. D. Naumann (ed.). *The insects of Australia: A textbook for students and research workers.* 2nd edn, vol. 1. Cornell University Press, Ithaca.

Kidd, N. A. C. 1988. The large pine aphid on Scots pine in Britain. pp. 111–128. In A. A. Berryman (ed.). *Dynamics of forest insect populations: Patterns, causes, implications.* Plenum Press, New York.

Kimberling, D. N., E. R. Scott, and P. W. Price. 1990. Testing a new hypothesis: Plant vigor and phylloxera distribution on wild grape in Arizona. *Oecologia* 84: 1–8.

Klimetzek, D. 1990. Population dynamics of pine-feeding insects: A historical study. pp. 3–10. In A. D. Watt, S. R. Leather, M. D. Hunter, and N. A. C. Kidd (eds.). *Population dynamics of forest insects.* Intercept, Andover, U.K.

Klomp, H. 1966. The dynamics of a field population of the pine looper, *Bupalus piniarus* L. (Lep., Geom.). *Adv. Ecol. Res.* 3: 207–305.

Kolehmainen, J., H. Roininen, R. Julkunen-Tiitto, and J. Tahvanainen. 1994. Importance of phenolic glucosides in host selection of shoot galling sawfly, *Euura amerinae*, on *Salix pentandra. J. Chem. Ecol.* 20: 2455–2466.

Kopelke, J.-P. 1982. Die gallenbildenden *Pontania*-Arten—ihre Sonderstellung unter den Blattwespen. Teil I: Gallenbildung, Entwicklung und Phänologie. *Natur und Museum* 112: 356–365.

Kopelke, J.-P. 1988. Zur Biologie und Ökologie der Arten des Brutparasiten-Parasitoiden–Komplexes von gallenbildenden Blattwespen der Gattung *Pontania* (Hymenoptera: Tenthredinidae: Nematinae). *Mitt. dtsch. Ges. allg. angew. Entomol.* 6: 150–155.

Kopelke, J.-P. 1994. Die Arten der *Pontania dolichura*-Gruppe in Mittel- und Nordeuropa (Insecta: Hymenoptera: Tenthredinidae: Nematinae). *Senckenbergiana Biol.* 74: 127–145.

Kopelke, J.-P. 2000. *Euura auritae* sp.n – ein neuer Gallenerzeuger der *atra*-Gruppe in Europa (Insecta, Hymenoptera, Tenthredinidae, Nematinae). *Senckenbergiana Biol.* 80: 159–163.

Kopelke, J.-P. 2001. Die Artengruppen von *Euura mucronata* und *E. laeta* in Europa (Insecta, Hymenoptera, Tenthredinidae, Nematinae). *Senckenbergiana Biol.* 81: 191–225.

Koricheva, J., S. Larsson, and E. Haukioja. 1998. Insect performance on experimentally stressed woody plants: A meta-analysis. *Annu. Rev. Entomol.* 43: 195–216.

Kozlov, M. V. In press. Density fluctuations of the leafminer *Phyllonorycter strigulatella* (Lepidoptera: Gracillariidae) in the imapct zone of a power plant. *Environ. Pollution.*

Kozlov, M. V., S. Koponen, J. Konki, P. Niemälä, and P. W. Price. In press. Larval food and feeding habit contribute to periodicity and magnitude of density fluctuations in subarctic forest moths.

Krebs, C. J. 1994. *Ecology: The experimental analysis of distribution and abundance.* 4th edn. HarperCollins, New York.

Krebs, C. J. 1995. Two paradigms of population regulation. *Wildlife Res.* 22: 1–10.

Krebs, C. J. 1999. *Ecological methodology.* 2nd edn. Addison-Wesley, Menlo Park.

Kuhn, T. S. 1962. *The structure of scientific revolutions.* University of Chicago Press, Chicago.

Kunin, W. E., and W. J. Gaston. 1993. The biology of rarity: Patterns, causes and consequences. *Trends Ecol. Evol.* 8: 298–301.

Lack, D. 1946. Clutch and brood size in the robin. *Brit. Birds* 39: 98–109, 130–135.

Lack, D. 1947a. The significance of clutch-size in the partridge (*Perdix perdix*). *J. Anim. Ecol.* 16: 19–25.

Lack, D. 1947b. The significance of clutch size. *Ibis* 89: 302–352.

Lack, D. 1948a. The significance of clutch size. *Ibis* 90: 25–45.

Lack, D. 1948b. Further notes on clutch and brood size in the robin. *Brit. Birds* 41: 98–104, 130–137.

Lack, D. 1948c. The significance of litter-size. *J. Anim. Ecol.* 17: 45–50.

Lack, D. 1954. *The natural regulation of animal numbers.* Oxford University Press, London.

Långström, B. 1980. Distribution of pine shoot beetle attacks within the crown of Scots pine. *Stadia Forestalia Suecica* 154: 1–25.

Larsson, S. 1989. Stressful times for the plant stress–insect performance hypothesis. *Oikos* 56: 277–283.

Larsson, S., C. Björkman, and N. A. C. Kidd. 1993. Outbreaks of diprionid sawflies: Why some species and not others? pp. 453–483. In M. Wagner and K. F. Raffa (eds.). *Sawfly life history adaptations to woody plants.* Academic Press, San Diego.

Lawton, J. H. 1983. Plant architecture and the diversity of phytophagous insects. *Annu. Rev. Entomol.* 28: 23–39.

Lawton, J. H. 1990. Species richness and population dynamics of animal assemblages: Patterns in body size: abundance space. *Phil. Trans. R. Soc. Lond. B* 330: 283–291.

Lawton, J. H. 1991. Species richness, population abundances, and body sizes in insect communities: Tropical versus temperate comparisons. pp. 71–89. In P. W. Price, T. M. Lewinsohn, G. W. Fernandes, and W. W. Benson (eds.). *Plant-animal interactions: Evolutionary ecology in tropical and temperate regions.* Wiley, New York.

Lawton, J. H. 1992. There are not 10 million kinds of population dynamics. *Oikos* 63: 337–338.

Lawton, J. H. 1999. Are there general laws in ecology? *Oikos* 84: 177–192.

Lawton, J. H. 2000. *Community ecology in a changing world.* Ecology Institute, Oldendorf/Luhe, Germany.

Lawton, J. H., and P. W. Price. 1979. Species richness of parasites on hosts: Agromyzid flies on the British Umbelliferae. *J. Anim. Ecol.* 48: 619–637.

Lawton, J. H., and D. Schroder. 1977. Effects of plant type, size of geographical range and taxonomic isolation on number of insect species associated with British plants. *Nature* 265: 137–140.

Lawton, J. H., and D. Schroder. 1978. Some observations on the structure of phytophagous insect communities: The implications for biological control. pp. 57–73. In *Proc. 4th Int. Symp. Biol. Control Weeds.* University of Florida Press, Gainesville.

Leyva, K. J., K. M. Clancy, and P. W. Price. 2000. Oviposition preference and larval performance of the western spruce budworm (Lepidoptera: Tortricidae). *Environ. Entomol.* 29: 281–289.

Leyva, K. J., K. M. Clancy, and P. W. Price. In press. Oviposition strategies employed by the western spruce budworm: Tests of predictions from the phylogenetic constraints hypothesis. *Agri. For. Entomol.*

Liebhold, A., and N. Kamata. 2000. Population dynamics of forest-defoliating insects. *Popul. Ecol.* 42: 205–209.

Liebhold, A., J. Elkinton, D. Williams, and R.-M. Muzika. 2000. What causes outbreaks of the gypsy moth in North America? *Popul. Ecol.* 42: 257–266.

Lightfoot, D. C., and W. G. Whitford. 1987. Variation in insect densities on desert creosote bush: Is nitrogen a factor? *Ecology* 68: 547–557.

Ligon, J. D. 1993. The role of phylogenetic history in the evolution of contemporary avian mating and parental care systems. *Current Ornithol.* 10: 1–46.

Lockwood, J. A. 2001. Voices from the past: What we can learn from the Rocky Mountain Locust. *Am. Entomol.* 47: 208–215.

Lockwood, J. A., and A. B. Ewen. 1997. Biological control of rangeland grasshoppers and locusts. pp. 421–442. In S. K. Gangwere, M. C. Muralirangan, and M. Muralirangen (eds.). *The bionomics of grasshoppers, katydids and their kin.* CAB International, Wallingford, U.K.

Lotka, A. J. 1924. *Elements of physical biology.* Williams and Wilkins, Baltimore.

Lumley, J. S. P., and W. Benjamin. 1994. *Research: Some ground rules.* Oxford University Press, Oxford, U.K.

MacArthur, R. H. 1962. Some generalized theorems of natural selection. *Proc. Natl Acad. Sci. U.S.A.* 48: 1893–1897.

MacArthur, R. H. 1972a. *Geographical ecology: Patterns in the distribution of species.* Harper and Row, New York.

MacArthur, R. H. 1972b. Coexistence of species. pp. 253–259. In J. Behnke (ed.). *Challenging biological problems.* Oxford University Press, New York.

MacArthur, R. H., and E. O. Wilson. 1967. *The theory of island biogeography.* Princeton University Press, Princeton.

Maddison, W. P. 1990. A method for testing the correlated evolution of two binary characters: Are gains or losses concentrated on certain branches of a phylogenetic tree? *Evolution* 44: 539–557.

Malthus, T. R. 1798. *An essay on the principle of population as it affects the future improvement of society.* Johnson, London.

Manson-Bahr, P. 1963. The story of malaria: The drama and the actors. *Int. Rev. Trop. Med.* 2: 329–390.

Marques, R. S. A., E. S. A. Marques, and P. W. Price. 1994. Female behavior and oviposition choices by an eruptive herbivore, *Disonycha pluriligata* (Coleoptera: Chrysomelidae). *Environ. Entomol.* 23: 887–892.

Martin, J. L. 1956. The bionomics of the aspen blotch miner, *Lithocolletis salicifoliella* Cham. (Lepidoptera: Gracillariidae). *Can. Entomol.* 88: 155–168.

Martins, E. P. (ed.). 1996. *Phylogenies and the comparative method in animal behavior.* Oxford University Press, Oxford, U.K.

Marzluff, J. M., and K. P. Dial. 1991. Life history correlates of taxonomic diversity. *Ecology* 72: 428–439.

Mason, R. R. 1987. Nonoutbreak species of forest Lepidoptera. pp. 31–57. In P. Barbosa and J. C. Schultz (eds.). *Insect outbreaks.* Academic Press, San Diego.

Mason, R. R., and B. E. Wickman. 1988. The Douglas-fir tussock moth in the interior Pacific Northwest. pp. 179–209. In A. A. Berryman (ed.). *Dynamics of forest insect populations: Patterns, causes, implications.* Plenum Press, New York.

Mattson, W. J. 1980. Cone resources and the ecology of the red pine cone beetle, *Conophthorus resinosae* (Coleoptera: Scolytidae). *Ann. Entomol. Soc. Am.* 73: 390–396.

Mattson, W. J., G. A. Simmons, and J. A. Witter. 1988. The spruce budworm in eastern North America. pp. 309–330. In A. A. Berryman (ed.). *Dynamics of forest insect populations: Patterns, causes, implications.* Plenum Press, New York.

Mattson, W. J., P. Niemalä, I. Millers, and Y. Inguanzo. 1994. *Immigrant phytophagous insects on woody plants in the United States and Canada: An annotated list.* U.S. Dept. Agr. For. Serv. N. Central For. Expt. St. Gen. Tech. Rep. NC-169: 1–27.

Mayr, E. 1961. Cause and effect in biology. *Science* 134: 1501–1506.

Mayr, E. 1982. *The growth of biological thought: Diversity, evolution, and inheritance.* Belknap Press of Harvard University Press, Cambridge.

McClure, M. S. 1988. The armored scales of hemlock. pp. 45–65. In A. A. Berryman (ed.). *Dynamics of forest insect populations: Patterns, causes, implications.* Plenum Press, New York.

McKitrick, M. C. 1993. Phylogenetic constraint in evolutionary theory: Has it any explanatory power? *Annu. Rev. Ecol. Syst.* 24: 307–330.

McLeod, J. M. 1970. The epidemiology of the Swaine jack-pine sawfly, *Neodiprion swainei* Midd. *Forest Chron.* 46: 126–133.

McLeod, J. M. 1972. The Swaine jack pine sawfly, *Neodiprion swainei*, life system: Evaluating the long-term effects of insecticide applications in Quebec. *Environ. Entomol.* 1: 371–381.

McPeek, M. A. 1995. Testing hypotheses about evolutionary change on single branches of a phylogeny using evolutionary contrasts. *Am. Nat.* 145: 686–703.

Meijden, E. van der, R. M. Nisbet, and M. J. Crawley. 1998. The dynamics of a herbivore–plant interaction, the cinnabar moth and ragwort. pp. 291–308. In J. P. Dempster and I. F. G. McLean (eds.). *Insect populations in theory and in practice.* Kluwer, Dordrecht, The Netherlands.

Mendonça, M. de S. 2001. Galling insect diversity patterns: The resource synchronization hypothesis. *Oikos* 95: 171–176.

Menges, E. S., and N. Kohfeldt. 1995. Life history strategies of Florida scrub plants in relation to fire. *Bull. Torrey Bot. Club* 122: 282–297.

Menzel, P., and F. D'Aluisio. 1998. *Man eating bugs: The art and science of eating insects.* Ten Speed Press, Berkeley.

Metcalf, R. L., and E. R. Metcalf. 1992. *Plant kairomones in insect ecology and control.* Chapman and Hall, New York.

Metcalf, R. L., and R. A. Metcalf. 1993. *Destructive and useful insects: Their habits and control.* 5th edn. McGraw-Hill, New York.

Miller, C. A. 1958. The measurement of spruce budworm populations and mortality during the first and second larval instars. *Can. J. Zool.* 36: 409–422.

Miller, C. A. 1963. The spruce budworm. pp. 12–19. In R. F. Morris (ed.). *The dynamics of epidemic spruce budworm populations.* Mem. Entomol. Soc. Can. 31. Entomological Society of Canada, Ottawa.

Miller, C. A. 1966. The black-headed budworm in eastern Canada. *Can. Entomol.* 98: 592–613.

Milne, A. 1957a. The natural control of insect populations. *Can. Entomol.* 89: 193–213.

Milne, A. 1957b. Theories of natural control of insect populations. *Cold Spring Harbor Symp. Quant. Biol.* 22: 253–267.

Milne, A. 1962. On a theory of natural control of insect population. *J. Theoret. Biol.* 3: 19–50.

Mitchell, E. R. 1979. Migration by *Spodoptera exigua* and *S. frugiperda*, North American style. pp. 386–393. In R. L. Rabb and G. G. Kennedy (eds.). *Movement of highly mobile insects: Concepts and methodology in research.* North Carolina State University Press, Raleigh.

Mitter, C., B. Farrell, and B. Wiegmann. 1988. The phylogenetic study of adaptive zones: Has phytophagy promoted insect diversification? *Am. Nat.* 132: 107–128.

Mitter, C., B. Farrell, and D. J. Futuyma. 1991. Phylogenetic studies of insect–plant interactions: Insights into the genesis of diversity. *Trends Ecol. Evol.* 6: 290–293.

Mook, L. J. 1963. Birds and the spruce budworm. pp. 268–271. In R. M. Morris (ed.). *The dynamics of epidemic spruce budworm populations.* Mem. Entomol. Soc. Can. 31. Entomological Society of Canada, Ottawa.

Mopper, S., and T. Whitham. 1986. Natural bonsai of Sunset Crater. *Nat. Hist.* 95: 42–47.

Morgan, F. D., and G. S. Taylor. 1988. The white lace lerp in southeastern Australia. pp. 129–140. In A. A. Berryman (ed.). *Dynamics of forest insect populations: Patterns, causes, implications.* Plenum Press, New York.

Morrill, W. L., J. W. Gabor, D. K. Weaver, G. D. Kushnak, and N. J. Irish. 2000. Effect of host plant quality on the sex ratio and fitness of female wheat stem sawflies (Hymenoptera: Cephidae). *Environ. Entomol.* 29: 195–199.

Morris, R. F. 1955. The development of sampling techniques for forest insect defoliators, with particular reference to the spruce budworm. *Can. J. Zool.* 33: 225–294.

Morris, R. F. 1959. Single-factor analysis in population dynamics. *Ecology* 40: 580–588.

Morris, R. F. 1960. Sampling insect populations. *Annu. Rev. Entomol.* 5: 243–264.

Morris, R. F. 1963a. Predictive population equations based on key factors. *Mem. Entomol. Soc. Can.* 32: 16–21.

Morris, R. F. (ed.). 1963b. *The dynamics of epidemic spruce budworm populations*. Mem. Entomol. Soc. Can. 31. Entomological Society of Canada, Ottawa.

Morris, R. F. 1969. Approaches to the study of population dynamics. pp. 9–28. In W. E. Waters (ed.). *Forest insect population dynamics*. U.S. Dept. Agr. For. Serv. Res. Paper NE-125. U.S. Government Printing Office, Washington, D.C.

Morris, R. F., and C. A. Miller. 1954. The development of life tables for the spruce budworm. *Can. J. Zool.* 32: 283–301.

Mott, D. G. 1963. The analysis of survival in small larvae in the unsprayed area. pp. 42–52. In R. F. Morris (ed.). *The dynamics of epidemic spruce budworm populations*. Mem. Entomol. Soc. Can. 31. Entomological Society of Canada, Ottawa.

Mulkey, S. S., R. L. Chazdon, and A. P. Smith (eds.). 1996. *Tropical forest plant ecophysiology*. Chapman and Hall, New York.

Munster-Swendsen, M. 1985. A simulation study of primary, clepto- and hyper-parasitism in *Epinotia tedella* (Cl.) (Lepidoptera: Tortricidae). *J. Anim. Ecol.* 54: 683–695.

Murdoch, W. W. 1994. Population regulation in theory and practice. *Ecology* 75: 271–287.

Myers, J. H. 1988. Can a general hypothesis explain population cycles of forest Lepidoptera? *Adv. Ecol. Res.* 18: 179–242.

Myers, J. H. 2000. Population fluctuations of the western tent caterpillar in southwestern British Columbia. *Popul. Ecol.* 42: 231–241.

Nealis, V. G., and P. V. Lomic. 1994. Host-plant influence on the population ecology of the jack pine budworm, *Choristoneura pinus* (Lepidoptera: Tortricidae). *Ecol. Entomol.* 19: 367–373.

Neilsen, M. M., and R. F. Morris. 1964. The regulation of European spruce sawfly numbers in the Maritime Provinces of Canada from 1937–1963. *Can. Entomol.* 96: 773–784.

Nicholson, A. J. 1933. The balance of animal populations. *J. Anim. Ecol.* 2: 132–178.

Nicholson, A. J. 1954. An outline of the dynamics of animal populations. *Austral. J. Zool.* 2: 9–65.

Nicholson, A. J. 1957. The self-adjustment of populations to change. *Cold Spring Harbor Symp. Quant. Biol.* 22: 153–172.

Nicholson, A. J., and V. A. Bailey. 1935. The balance of animal populations. Part I. *Proc. Zool. Soc. London* 1935: 551–598.

Niemalä, P., and E. Haukioja. 1982. Seasonal patterns in species richness of herbivores: Macrolepidopteran larvae on Finnish deciduous trees. *Ecol. Entomol.* 7: 169–175.

Niklas, K. J., and T. D. O'Rourke. 1982. Growth patterns of plants that maximize vertical growth and minimize internal stresses. *Am. J. Bot.* 69: 1367–1375.

Nordlund, D. A., R. L. Jones, and W. J. Lewis (eds.). 1981. *Semiochemicals: Their role in pest control*. Wiley, New York.

Nothnagle, P. J., and J. C. Schultz. 1987. What is a forest pest? pp. 59–80. In P. Barbosa and J. C. Schultz (eds.). *Insect outbreaks*. Academic Press, San Diego.

Nyman, T. 2000a. Introduction. pp. 7–39. In T. Nyman (ed.). Phylogeny and ecological evolution of gall-inducing sawflies (Hymenoptera: Tenthredinidae). PhD thesis, University of Joensuu, Joensuu, Finland.

Nyman, T. 2000b. The willow bud galler *Euura mucronata* Hartig (Hymenoptera: Tenthredinidae): One polyphage or many monophages? Article III, pp. 1–12. In T. Nyman (ed.). Phylogeny and ecological evolution of gall-inducing sawflies (Hymenoptera: Tenthredinidae). PhD thesis, University of Joensuu, Joensuu, Finland.

Nyman, T., H. Roininen, and J. A. Vuorinen. 1998. Evolution of different gall types in willow-feeding sawflies (Hymenoptera: Tenthredinidae). *Evolution* 52: 465–474.

Nyman, T., A. Widmer, and H. Roininen. 2000. Evolution of gall morphology and host-plant relationships in willow-feeding sawflies (Hymenoptera: Tenthredinidae). *Evolution* 54: 526–533.

Odum, E. P. 1959. *Fundamentals of ecology.* Saunders, Philadelphia.

Okuda, S., and J. Yukawa. 2000. Life history strategy of *Tokiwadiplosis matecola* (Diptera: Cecidomyiidae) relying upon the lammas shoots of *Lithocarpus edulis* (Fagaceae). *Entomol. Sci.* 3: 47–56.

Ohgushi, T. 1995. Adaptive behavior produces stability in herbivorous lady beetle populations. pp. 303–319. In N. Cappuccino and P. W. Price (eds.). *Population dynamics: New approaches and synthesis.* Academic Press, San Diego.

Orians, G. H. 1962. Natural selection and ecological theory. *Am. Nat.* 96: 257–263.

Paine, T. D., K. F. Raffa, and T. C. Harrington. 1997. Interactions among scolytid bark beetles, their associated fungi, and live host conifers. *Annu. Rev. Entomol.* 42: 179–206.

Pearl, R., and J. R. Miner. 1935. Experimental studies on the duration of life. XIV. The comparative mortality of certain lower organisms. *Q. Rev. Biol.* 10: 60–79.

Pearl, R., and S. L. Parker. 1921. Experimental studies on the duration of life. I. Introductory discussion of the duration of life in *Drosophila. Am. Nat.* 55: 481–509.

Pearl, R., and L. J. Reed. 1920. On the rate of growth of the population of the United States since 1790 and its mathematical representation. *Proc. Natl Acad. Sci. U.S.A.* 6: 275–288.

Peters, R. H. 1983. *The ecological implications of body size.* Cambridge University Press, Cambridge, U.K.

Pfadt, R. E. 1988. *Field guide to common western grasshoppers.* U.S. Dept. Agr., APHIS Wyoming Expt. Stat. Bull. 912.

Pickett, S. T. A., and P. S. White (eds.). 1985. *The ecology of natural disturbance and patch dynamics.* Academic Press, San Diego.

Pickett, S. T. A., J. Kolasa, and C. G. Jones. 1994. *Ecological understanding: The nature of theory and the theory of nature.* Academic Press, San Diego.

Pierce, W. D., R. A. Cushman, and C. E. Hood. 1912. The insect enemies of the cotton boll weevil. *U.S. Dept. Agr. Bur. Entomol. Bull.* 100: 1–99.

Pimm, S. L. 1984. Food chains and return times. pp. 397–412. In D. R. Strong, D. Simberloff, L. G. Abele, and A. B. Thistle (eds.). *Ecological communities: Conceptual issues and the evidence.* Princeton University Press, Princeton.

Pires, C. S. S. 1998. Influence of the host plant on the population dynamics of the spittlebug *Deois flavopicta.* PhD dissertation, Northern Arizona University, Flagstaff.

Pires, C. S. S., and P. W. Price. 2000. Patterns of host plant growth and attack and establishment of gall-inducing wasp (Hymenoptera: Cynipidae). *Environ. Entomol.* 29: 49–54.

Pires, C. S. S., P. W. Price, and E. G. Fontes. 2000. Preference–performance linkage in the neotropical spittlebug *Deois flavopicta,* and its relation to the phylogenetic constraints hypothesis. *Ecol. Entomol.* 25: 71–80.

Platt, J. R. 1964. Strong inference. *Science* 146: 347–353.

Pollard, E. 1979. Population ecology and change in range of the white admiral butterfly *Ladoga camilla* in England, U.K. *Ecol. Entomol.* 4: 61–74.

Prada, M., O. J. Marini-Filho, and P. W. Price. 1995. Insects in flower heads of *Aspilia foliacea* (Asteraceae) after a fire in a central Brazilian savanna: Evidence for the plant vigor hypothesis. *Biotropica* 27: 513–518.

Preszler, R. W., and P. W. Price. 1988. Host quality and sawfly populations: A new approach to life table analysis. *Ecology* 69: 2012–2020.

Preszler, R. W., and P. W. Price. 1995. A test of plant-vigor, plant stress, and plant-genotype effects on leaf-miner oviposition and performance. *Oikos* 74: 485–492.

Price, P. W. 1973. Reproductive strategies in parasitoid wasps. *Am. Nat.* 107: 684–693.

Price, P. W. 1974. Strategies for egg production. *Evolution* 28: 76–84.

Price, P. W. 1975. Reproductive strategies of parasitoids. p. 87–111. In P. W. Price (ed.). *Evolutionary strategies of parasitic insects and mites.* Plenum Press, New York.

Price, P. W. 1980. *Evolutionary biology of parasites.* Princeton University Press, Princeton.

Price, P. W. 1988. Inversely density-dependent parasitism: The role of plant refuges for hosts. *J. Anim. Ecol.* 57: 89–96.

Price, P. W. 1989. Clonal development of coyote willow, *Salix exiguae* (Salicaceae), and attack by the shoot-galling sawfly, *Euura exiguae* (Hymenoptera: Tenthredinidae). *Environ. Entomol.* 18: 61–68.

Price, P. W. 1990a. Evaluating the role of natural enemies in latent and eruptive species: New approaches to life table construction. pp. 221–232. In A. D. Watt, S. R. Leather, M. D. Hunter, and N. A. C. Kidd (eds.). *Population dynamics of forest insects.* Intercept, Andover, U.K.

Price, P. W. 1990b. Insect herbivore population dynamics: Is a new paradigm available? *Symp. Biol. Hung.* 39: 177–190.

Price, P. W. 1991a. Darwinian methodology and the theory of insect herbivore population dynamics. *Ann. Entomol. Soc. Am.* 84: 465–473.

Price, P. W. 1991b. Evolutionary theory of host and parasitoid interactions. *Biol. Control.* 1: 83–93.

Price, P. W. 1991c. The plant vigor hypothesis and herbivore attack. *Oikos* 62: 244–251.

Price, P. W. 1991d. Patterns in communities along latitudinal gradients. pp. 51–69. In P. W. Price, T. M. Lewinsohn, G. W. Fernandes, and W. W. Benson (eds.). *Plant-animal interactions: Evolutionary ecology in tropical and temperate regions.* Wiley, New York.

Price, P. W. 1992a. Evolution and ecology of gall-inducing sawflies. pp. 208–224. In J. D. Shorthouse and O. Rohfritsch (eds.). *Biology of insect-induced galls.* Oxford University Press, New York.

Price, P. W. 1992b. Plant resources as a mechanistic basis for insect herbivore population dynamics. pp. 139–173. In M. D. Hunter, T. Ohgushi, and P. W. Price (eds.). *Effects of resource distribution on animal–plant interactions.* Academic Press, San Diego.

Price, P. W. 1994a. Evolution of parasitoid communities. pp. 472–491. In B. A. Hawkins and W. Sheehan (eds.). *Parasitoid community ecology.* Oxford University Press, Oxford, U.K.

Price, P. W. 1994b. Phylogenetic constraints, adaptive syndromes, and emergent properties: From individuals to population dynamics. *Res. Popul. Ecol.* 36: 3–14.

Price, P. W. 1997. *Insect ecology.* 3rd edn. Wiley, New York.

Price, P. W. 2002. Resource-driven terrestrial interaction webs. *Ecol. Res.* 17: 241–247.

Price, P. W., and T. G. Carr. 2000. Comparative ecology of membracids and tenthredinids in a macroevolutionary context. *Evol. Ecol. Res.* 2: 645–665.

Price, P. W., and K. M. Clancy. 1986a. Multiple effects of precipitation on *Salix lasiolepis* and populations of the stem-galling sawfly, *Euura lasiolepis*. *Ecol. Res.* 1: 1–14.

Price, P. W., and K. M. Clancy. 1986b. Interactions among three trophic levels: Gall size and parasitoid attack. *Ecology* 67: 1593–1600.

Price, P. W., and T. P. Craig. 1984. Life history, phenology, and survivorship of a stem-galling sawfly, *Euura lasiolepis* (Hymenoptera: Tenthredinidae), on the arroyo willow, *Salix lasiolepis*, in northern Arizona. *Ann. Entomol. Soc. Am.* 77: 712–719.

Price, P. W., and D. Gerling. In press. Complex architecture of *Tamarix nilotica* and resource utilization by the spindle-gall moth *Amblypalpis olivierella* (Lepidoptera: Gelechiidae). *Israel J. Entomol.*

Price, P. W., and T. Ohgushi. 1995. Preference and performance linkage in a *Phyllocolpa* sawfly on the willow, *Salix miyabeana*, on Hokkaido. *Res. Popul. Ecol.* 37: 23–28.

Price, P. W., and H. Pschorn-Walcher. 1988. Are galling insects better protected against parasitoids than exposed feeders? A test using tenthredinid sawflies. *Ecol. Entomol.* 13: 195–205.

Price, P. W., and H. Roininen. 1993. The adaptive radiation in gall induction. pp. 229–257. In M. R. Wagner and K. F. Raffa (eds.). *Sawfly life history adaptations to woody plants*. Academic Press, San Diego.

Price, P. W., C. E. Bouton, P. Gross, B. A. McPherson, J. N. Thompson, and A. E. Weis. 1980. Interactions among three trophic levels: Influence of plants on interactions between insect herbivores and natural enemies. *Annu. Rev. Ecol. Syst.* 11: 41–65.

Price, P. W., G. W. Fernandes, and G. L. Waring. 1987a. Adaptive nature of insect galls. *Environ. Entomol.* 16: 15–24.

Price, P. W., H. Roininen, and J. Tahvanainen. 1987b. Plant age and attack by the bud galler, *Euura mucronata*. *Oecologia* 73: 334–337.

Price, P. W., H. Roininen, and J. Tahvanainen. 1987c. Why does the bud-galling sawfly, *Euura mucronata*, attack long shoots? *Oecologia* 74: 1–6.

Price, P. W., G. L. Waring, R. Julkunen-Tiitto, J. Tahvanainen, H. A. Mooney, and T. P. Craig. 1989. The carbon–nutrient balance hypothesis in within species phytochemical variation of *Salix lasiolepis*. *J. Chem. Ecol.* 15: 1117–1131.

Price, P. W., N. Cobb, T. P. Craig, G. W. Fernandes, J. K. Itami, S. Mopper, and R. W. Preszler. 1990. Insect herbivore population dynamics on trees and shrubs: New approaches relevant to latent and eruptive species and life table development. pp. 1–38. In E. A. Bernays (ed.). *Insect–plant interactions*, vol. 2. CRC Press, Boca Raton.

Price, P. W., K. M. Clancy, and H. Roininen. 1994. Comparative population dynamics of the galling sawflies. pp. 1–11. In P. W. Price, W. J. Mattson, and Y. N. Baranchikov (eds.). *Ecology and evolution of gall-forming insects*. U.S. Dept. Agr. For. Serv. N. Central For. Expt. Sta. Gen. Tech. Rep. NC-174.

Price, P. W., T. P. Craig, and H. Roininen. 1995a. Working toward theory on galling sawfly population dynamics. pp. 321–338. In N. Cappuccino and P. W. Price (eds.). *Population dynamics: New approaches and synthesis*. Academic Press, San Diego.

Price, P. W., I. Andrade, C. Pires, E. Sujii, and E. M. Vieira. 1995b. Gradient analysis using plant modular structure: Pattern in plant architecture and insect herbivore utilization. *Environ. Entomol.* 24: 497–505.

Price, P. W., H. Roininen, and J. Tahvanainen. 1997. Willow tree shoot module length and the attack and survival pattern of a shoot-galling sawfly, *Euura atra* (Hymenoptera: Tenthredinidae). *Entomol. Fennica* 8: 113–119.

Price, P. W., T. P. Craig, and M. D. Hunter. 1998a. Population ecology of a gall-inducing sawfly, *Euura lasiolepis*, and relatives. pp. 323–340. In J. P. Dempster and I. F. G. McLean (eds.). *Insect populations: In theory and in practice*. Kluwer, Dordrecht, The Netherlands.

Price, P. W., G. W. Fernandes, A. C. F. Lara, J. Brawn, H. Barrios, M. G. Wright, S. P. Ribeiro, and N. Rothcliff. 1998b. Global patterns in local number of insect galling species. *J. Biogeog.* 25: 581–591.

Price, P. W., H. Roininen, and A. Zinovjev. 1998c. Adaptive radiation of gall-inducing sawflies in relation to architecture and geographic range of willow host plants. pp. 196–203. In G. Csóka, W. J. Mattson, G. N. Stone, and P. W. Price (eds.). *Biology of gall-inducing arthropods*. U.S. Dept. Agr. For. Serv. N. Central Res. Sta. Gen. Tech. Rep. NC-199.

Price, P. W., H. Roininen, and T. Ohgushi. 1999. Comparative plant–herbivore interactions involving willows and three gall-inducing sawfly species in the genus *Pontania* (Hymenoptera: Tenthredinidae). *Ecoscience* 6: 41–50.

Promislow, D. E. L. 1996. Using comparative approaches to integrate behavior and population biology. pp. 288–323. In E. P. Martins (ed.). *Phylogenies and the comparative method in animal behavior*. Oxford University Press, Oxford, U.K.

Pschorn-Walcher, H. 1982. Unterordnung Symphyta, Pflanzenwespen. pp. 4–196. In W. Schwenke (ed.). *Die Forstschädlinge Europas*, vol. 4, *Hautflügler und Zweiflägler*. Paul Perey Verlag, Hamburg, Germany.

Quintana-Ascencio, P. F., and E. S. Menges. 1996. Inferring metapopulation dynamics from patch-level incidence of Florida scrub plants. *Conservation Biol.* 10: 1210–1219.

Quintana-Ascencio, P. F., R. W. Dolan, and E. S. Menges. 1998. *Hypericum cumulicola* demography in unoccupied and occupied Florida scrub patches with different time-since-fire. *J. Ecol.* 86: 640–651.

Rabinowitz, D. 1981. Seven forms of rarity. pp. 205–217. In H. Synge (ed.). *The biological aspects of rare plant conservation*. Wiley, New York.

Radtkey, R. R., and M. C. Singer. 1995. Repeated reversals of host-preference evolution in a specialist insect herbivore. *Evolution* 49: 351–359.

Raffa, K. F. 1988. The mountain pine beetle in western North America. pp. 505–530. In A. A. Berryman (ed.). *Dynamics of forest insect populations: Patterns, causes, implications*. Plenum Press, New York.

Rainey, R. C. 1979. Interactions between weather systems and populations of locusts and noctuids in Africa. pp. 109–119. In R. L. Rabb and G. G. Kennedy (eds.). *Movement of highly mobile insects: Concepts and methodology in research*. North Carolina State University Press, Raleigh.

Rasch, C., and H. Rembold. 1994. Carbon dioxide: Highly attractive signal for larvae of *Helicoverpa armigera*. *Naturwissenschaften* 81: 228–229.

Raulo, J., and M. Leikola. 1974. Tutkimuksia puiden vuotuisen pituuskasvon ajoittumisesta. *Communicationes Inst. For. Fenniae* 81(2): 1–19.

Raunkiaer, C. 1934. *The life forms of plants and statistical plant geography*. Clarendon, Oxford, U.K.

Rausher, M. D. 1982. Population differentiation in *Euphydryas editha* butterflies: Larval adaptation to different hosts. *Evolution* 36: 581–590.

Readshaw, J. L. 1965. A theory of phasmatid outbreak release. *Austral. J. Zool.* 13: 475–490.

Real, L. (ed.). 1983. *Pollination biology*. Academic Press, San Diego.

Real, L. A. 1992. Introduction to the symposium (Behavioral mechanisms in evolutionary biology). *Am. Nat.* 140: S1–S4.

Redfern, M., and R. A.. D. Cameron. 1998. The yew gall midge *Taxomyia taxi*: 28 years of interaction with chalcid parasitoids. pp. 90–105. In G. Csóka, W. J. Mattson, G. N. Stone, and P. W. Price (eds.). *The biology of gall-inducing arthropods*. U.S. Dept. Agr. For. Serv. N. Central Res. Sta. Gen. Tech. Rep. NC-199.

Reilly, S. M., and P. C. Wainwright. 1994. Conclusion: Ecological morphology and the power of integration. pp. 339–354. In P. C. Wainwright and S. M. Reilly (eds.). *Ecological morphology: Integrative organismal biology*. University of Chicago Press, Chicago.

Rentz, D. C. F. 1991. Orthoptera (grasshoppers, locusts, katydids, crickets). pp. 369–393. In I. D. Naumann (ed.). *The insects of Australia: A textbook for students and research workers*. 2nd edn, vol. 1. Cornell University Press, Ithaca.

Rentz, D. C. F. 1996. *Grasshopper country: The abundant orthopterid insects of Australia*. University of New South Wales Press, Sydney.

Reynoldson, T. B. 1957. Population fluctuations in *Urceolaria mitra* (Peritricha) and *Enchytraeus albidus* (Oligochaeta) and their bearing on regulation. *Cold Spring Harbor Symp. Quant. Biol.* 22: 313–324.

Rhodes, O. E., R. K. Chesser, and M. H. Smith (eds.). 1996. *Population dynamics in ecological space and time*. University of Chicago Press, Chicago.

Ricklefs, R. E. 1997. *The economy of nature*. 4th edn. Freeman, New York.

Ricklefs, R. E. 2000a. Density dependence, evolutionary optimization, and the diversification of avian life histories. *Condor* 102: 9–22.

Ricklefs, R. E. 2000b. Lack, Skutch, and Moreau: The early development of life-history thinking. *Condor* 102: 3–8.

Ricklefs, R. E., and D. B. Miles. 1994. Ecological and evolutionary inferences from morphology: An ecological perspective. pp. 13–41. In P. C. Wainwright and S. M. Reilly (eds.). *Ecological morphology: Integrative organismal biology*. University of Chicago Press, Chicago.

Ricklefs, R. E., and G. L. Miller. 2000. *Ecology*. 4th edn. Freeman, New York.

Ricklefs, R. E., J. M. Starck, and M. Konarzewski. 1998. Internal constraints on growth in birds. pp. 266–287. In J. M. Starck and R. E. Ricklefs (eds.). *Avian growth and development*. Oxford University Press, Oxford, U.K.

Riegert, P. W., A. B. Ewen, and J. A. Lockwood. 1997. A history of chemical control of grasshoppers and locusts 1940–1990. pp. 385–405. In S. K. Gangwere, M. C. Muralirangan, and M. Muralirangen (eds.). *The bionomics of grasshoppers, katydids and their kin*. CAB International, Wallingford, U.K.

Roininen, H. 1991. The ecology and evolution of the host plant relationships among willow-feeding sawflies. PhD thesis, University of Joensuu, Joensuu, Finland.

Roininen, H., P. W. Price, and J. Tahvanainen. 1993a. Colonization and extinction in a population of the shoot-galling sawfly, *Euura amerinae*. *Oikos* 68: 448–454.

Roininen, H., J. Vuorinen, J. Tahvanainen, and R. Julkunen-Tiitto. 1993b. Host preference and allozyme differentiation in shoot galling sawfly, *Euura atra*. *Evolution* 47: 300–308.

Roininen, H., P. W. Price, and J. P. Bryant. 1997. Response of galling insects to natural browsing by mammals in Alaska. *Oikos* 80: 481–484.

Roininen, H., P. W. Price, R. Julkunen-Tiitto, J. Tahvanainen, and A. Ikonen. 1999. Oviposition stimulant for a gall-inducing sawfly, *Euura lasiolepis*, on willow is a phenolic glucoside. *J. Chem. Ecol.* 25: 943–953.

Roininen, H., T. Carr, and P. W. Price. In press. Plant vigor hypothesis and the shoot galling fly, *Hexomyza schineri* (Diptera: Agromyzidae). *Environ. Entomol.*

Root, R. B. 1996. Maintaining ecology's broad scope. *Ecology* 77: 1311–1312.

Root, R. B., and S. J. Chaplin. 1976. The life styles of tropical milkweed bugs, *Oncopeltus* (Hemiptera: Lygaeidae) utilizing the same hosts. *Ecology* 57: 132–140.

Roques, A. 1988. The larch cone fly in the French Alps. pp. 1–28. In A. A. Berryman (ed.). *Dynamics of forest insect populations: Patterns, causes, implications.* Plenum Press, New York.

Rose, D. J. W. 1975. Field development and quality changes in successive generations of *Spodoptera exempta* (Wlk.), the African armyworm. *J. Appl. Ecol.* 21: 729–739.

Rose, D. J. W., and S. Khasimuddin. 1979. Wide-area monitoring of the African armyworm, *Spodoptera exempta* (Walker); (Lepidoptera: Noctuidae). pp. 212–219. In R. L. Rabb and G. G. Kennedy (eds.). *Movement of highly mobile insects: Concepts and methodology in research.* North Carolina State University Press, Raleigh.

Rosenthal, G. A., and M. R. Berenbaum (eds.). 1991. *Herbivores: Their interactions with secondary plant metabolites,* vol. 1, *The chemical participants.* 2nd edn. Academic Press, San Diego.

Rosenthal, G. A., and M. R. Berenbaum (eds.). 1992. *Herbivores: Their interactions with secondary plant metabolites,* vol. 2, *Ecological and evolutionary processes.* 2nd edn. Academic Press, San Diego.

Rosenthal, G. A., and D. H. Janzen (eds.). 1979. *Herbivores: Their interaction with secondary plant metabolites.* Academic Press, San Diego.

Royama, T. 1984. Population dynamics of the spruce budworm *Choristoneura fumiferana. Ecol. Monogr.* 54: 429–462.

Royama, T. 1992. *Analytical population dynamics.* Chapman and Hall, London.

Royama, T. 1996. A fundamental problem in key factor analysis. *Ecology* 77: 87–93.

Ruel, J., and T. G. Whitham. In press. Yesterday's most vigorous pinyon pines are today's moth eaten shrubs: Tree-rings predict herbivory. *Ecology.*

Ruohomäki, K., M. Tanhuanpää, M. P. Ayres, P. Kaitaniemi, T. Tammaru, and E. Haukioja. 2000. Causes of cyclicity of *Epirrita autumnata* (Lepidoptera: Geometridae): Grandiose theory and tedious practice. *Popul. Ecol.* 42: 211–223.

Rutowski, R. L. In press. Visual ecology of adult butterflies. In C. Boggs and W. Watt (eds.). *Ecology and evolution taking flights: Butterflies as model study systems.* University of Chicago Press, Chicago.

Sacchi, C. F., and P. W. Price. 1992. The relative roles of abiotic and biotic factors in seedling demography of arroyo willow (*Salix lasiolepis:* Salicaceae). *Am. J. Bot.* 79: 395–405.

Salt, G. W. 1979. A comment on the use of the term *emergent properties. Am. Nat.* 113: 145–161.

Samways, M. J. 1994. *Insect conservation biology.* Chapman and Hall, London.

Schaal, B. A. 1984. Life-history variation, natural selection, and maternal effects in plant populations. pp. 188–211. In R. Dirzo, and J. Sarukhan (eds.). *Perspectives on plant population ecology.* Sinauer, Sunderland.

Scholtz, C. H., and E. Holm. 1985. *Insects of southern Africa.* Butterworths, Durban, South Africa.

Schwerdtfeger, F. 1935. Studien über den Massenwechsel einiger Forstschädlinge. *Z. Forst- und Jagdwes.* 67: 15–38, 95–104, 449–482, 513–540.

Schwerdtfeger, F. 1941. Uber die Ursachen des Massenwechsels der Insekten. *Z. angew. Entomol.* 28: 254–303.

Sequeira, A. A., B. B. Normark, and B. D. Farrell. 2000. Evolutionary assembly of the conifer fauna: Distinguishing ancient from recent associations in bark beetles. *Proc. R. Soc. Lond. B* 267: 2359–2366.

Seyffarth, J. A. S., A. M. Caldero, and P. W. Price. 1996. Leaf rollers in *Ouratea hexasperma* (Ochnaceae): Fire effect and the plant vigor hypothesis. *Rev. Bras. Biol.* 56: 135–137.

Shapiro, A. M. 1970. The role of sexual behavior in density-related dispersal of pierid butterflies. *Am. Nat.* 104: 367–372.

Shapiro, A. M. 1981. The pierid red-egg syndrome. *Am. Nat.* 117: 276–294.

Shiga, M. 1979. Population dynamics of *Malacosoma neustria testacea* (Lepidoptera, Lasiocampidae). *Bull. Fruit Tree Res. Sta. A* 6: 59–168.

Showler, A. T. 1995. Locust (Orthoptera: Acrididae) outbreak in Africa and Asia, 1992–1994: An overview. *Am. Entomol.* 41: 179–185.

Sih, A., G. Englund, and O. Wooster. 1998. Emergent impacts of multiple predators on prey. *Trends Ecol. Evol.* 13: 350–355.

Simberloff, D. 1980. The sick science of ecology: Symptoms, diagnosis, and prescription. *Eidema* 1: 49–54.

Sinclair, A. R. E. 1970. Studies of the ecology of the East African buffalo. PhD thesis, Oxford University, Oxford, U.K.

Sinclair, A. R. E. 1973. Regulation, and population models for a tropical ruminant. *E. Afr. Wildlife J.* 11: 307–316.

Singer, M. C. 1971. Evolution of food-plant preference in the butterfly *Euphydryas editha*. *Evolution* 25: 383–389.

Singer, M. C. 1972. Complex components of habitat suitability within a butterfly colony. *Science* 176: 75–77.

Singer, M. C., and C. D. Thomas. 1996. Evolutionary responses of a butterfly metapopulation to human and climate-caused environmental variation. *Am. Nat.* 148: S9–S39.

Singer, M. C., D. Ng, and C. D. Thomas. 1988. Heritability of oviposition preference and its relationship to offspring performance within a single insect population. *Evolution* 42: 977–985.

Singer, M. C., D. Ng, D. Vasco, and C. D. Thomas. 1992. Rapidly evolving associations among oviposition preferences fail to constrain evolution of insect diet. *Am. Nat.* 139: 9–20.

Singer, M. C., C. D. Thomas, and C. Parmesan. 1993. Rapid human–induced evolution of insect–host associations. *Nature* 366: 681–683.

Skaife, S. H., J. Ledger, and A. Bannister. 1979. *African insect life.* Rev. edn. Struik, Cape Town, South Africa.

Skvortsov, A. K. 1968. *Willows of the U.S.S.R.: A taxonomical and geographical revision.* Moskovsoe obschchestvo ispytatelei prirody, Moscow. (in Russian)

Skvortsov, A. K. 1999. *Willows of Russia and adjacent countries: Taxonomical and geographical revision.* University of Joensuu, Finland.

Smith, D. R. 1979. Symphyta. pp. 3–137. In K. V. Krombein, P. D. Hurd, D. R. Smith, and B. D. Burks (eds.). *Catalog of Hymenoptera in America north of Mexico*, vol. 1, *Symphyta and Apocrita (Parasitica).* Smithsonian Institution, Washington, D. C.

Smith, D. R. 1993. Systematics, life history, and distribution of sawflies. In M. Wagner and K. F. Raffa (eds.). *Sawfly life history adaptations to woody plants.* Academic Press, San Diego.

Smith, E. L. 1968. Biosystematics and morphology of Symphyta. I. Stem-galling *Euura* of the California region, and a new female genitalic nomenclature. *Ann. Entomol. Soc. Am.* 61: 1389–1407.

Smith, H. S. 1935. The role of biotic factors in the determination of population density. *J. Econ. Entomol.* 28: 873–898.

Smith, T., and F. L. Kilbourne. 1893. *Investigations into the nature, causation, and prevention of Texas or southern cattle fever.* U.S. Dept. Agr. Bur. Anim. Indust. Bull. 1.

Snodgrass, R. E. 1935. *Principles of insect morphology.* McGraw-Hill, New York.

Solbreck, D. 1995. Long-term population dynamics of a seed-feeding insect in a landscape perspective. pp. 279–301. In N. Cappuccino and P. W. Price (eds.). *Population dynamics: New approaches and synthesis.* Academic Press, San Diego.

Southwood, T. R. E. (ed.). 1968. *Insect abundance.* Symp. R. Entomol. Soc. London 4. Royal Entomological Society of London, London.

Southwood, T. R. E. 1975. The dynamics of insect populations. pp. 151–199. In D. Pimentel (ed.). *Insects, science, and society.* Academic Press, San Diego.

Southwood, T. R. E., and H. N. Comins. 1976. A synoptic population model. *J. Anim. Ecol.* 45: 949–965.

Southwood, T. R. E., and P. M. Reader. 1976. Population census data and key factor analysis for the viburnum whitefly, *Aleurotrachelus jelenekii* (Frauenf.), on three bushes. *J. Anim. Ecol.* 45: 313–325.

Spencer, K. C. 1988. Chemical mediation of coevolution in the *Passiflora-Heliconius* interaction. pp. 167–240. In K. C. Spencer (ed.). *Chemical mediation of coevolution.* Academic Press, San Diego.

Spiegel, L. H., and P. W. Price. 1996. Plant aging and the distribution of *Rhyacionia neomexicana* (Lepidoptera: Tortricidae). *Environ. Entomol.* 25: 359–365.

Stange, G., J. Monro, S. Stowe, and C. B. Osmond. 1995. The CO_2 sense of the moth *Cactoblastis cactorum* and its probable role in the biological control of the CAM plant *Opuntia stricta. Oecologia* 102: 341–352.

Starck, J. M., and R. E. Ricklefs. 1998. *Avian growth and development.* Oxford University Press, New York.

Stebbins, G. L. 1950. *Variation and evolution in plants.* Columbia University Press, New York.

Steffan-Dewenter, I., and T. Tscharntke. 1997. Early succession of butterfly and plant communities on set-aside fields. *Oecologia* 109: 294–302.

Stein, S. J., and P. W. Price. 1995. Relative effects of plant resistance and natural enemies by plant developmental age on sawfly (Hymenoptera: Tenthredinidae) preference and performance. *Environ. Entomol.* 24: 909–916.

Stein, S. J., P. W. Price, W. G. Abrahamson, and C. F. Sacchi. 1992. The effect of fire on stimulating willow regrowth and subsequent attack by grasshoppers and elk. *Oikos* 65: 190–196.

Stein, S. J., P. W. Price, T. P. Craig, and J. K. Itami. 1994. Dispersal of a galling sawfly: Implications for studies of insect population dynamics. *J. Anim. Ecol.* 63: 666–676.

Stiling, P. D. 1998. *Ecology: Theories and applications.* 3rd edn. Prentice Hall, Upper Saddle River.

Story, R. N., W. H. Robinson, R. L. Pienkowski, and L. T. Kok. 1979. The biology and immature stages of *Taphrocerus schaefferi*, a leaf-miner of yellow nutsedge. *Ann. Entomol. Soc. Am.* 72: 93–98.

Strong, D. R., and D. A. Levin. 1979. Species richness of plant parasites and growth form of their hosts. *Am. Nat.* 114: 1–22.

Strong, D. R., J. H. Lawton, and T. R. E. Southwood. 1984. *Insects on plants: Community patterns and mechanisms.* Harvard University Press, Cambridge.

Sutcliffe, O. L., C. D. Thomas, T. J. Yates, and J. N. Greatorex-Davies. 1997. Correlated extinctions, colonizations and population fluctuations in a highly correlated ringlet butterfly metapopulation. *Oecologia* 109: 235–241.

Swetnam, T. W., and J. L. Betancourt. 1998. Mesoscale disturbance and ecological response to decadal climatic variability in the American Southwest. *J. Climate* 11: 3128–3147.

Swetnam, T. W., and A. M. Lynch. 1993. Multicentury, regional-scale patterns of western spruce budworm outbreaks. *Ecol. Monogr.* 63: 399–424.

Tamarin, R. H. (ed.). 1978. *Population regulation*. Dowden, Hutchinson and Ross, Stroudsburg.

Tenow, O. 1972. The outbreaks of *Oporinia autumnata* Bkh. and *Operophtera* spp. (Lep., Geometridae) in the Scandinavian mountain chain and northern Finland 1862–1968. *Zool. Bijdr. Upps.* Suppl. 2: 1–107.

Thomas, J. A. 1991. Rare species conservation: Case studies of European butter-flies. pp. 149–197. In I. F. Spellerberg, F. B. Goldsmith and M. G. Morris (eds.). *The scientific management of temperate communities for conservation*. Blackwell Scientific Publications, Oxford, U.K.

Thomas, J. A., R. T. Clarke, G. W. Elmes, and M. E. Hochberg. 1998. Population dynamics in the genus *Maculinea* (Lepidoptera: Lycaenidae). pp. 261–290. In J. P. Dempster and I. F. G. McLean (eds.). *Insect populations: In theory and in practice*. Kluwer, Dordrecht, The Netherlands.

Thomas, J. H. 1974. Factors influencing the numbers and distribution of the brown hairstreak, *Thecla betulae* L, and the black hairstreak, *Strymonidia pruni* L (Lepidoptera, Lycaenidae). PhD thesis, University of Leicester, Leicester, U.K.

Thompson, J. N., and O. Pellmyr. 1991. Evolution of oviposition behavior and host preference in Lepidoptera. *Annu. Rev. Entomol.* 36: 65–89.

Thompson, W. R. 1929. On natural control. *Parasitology* 21: 269–281.

Thompson, W. R. 1939. Biological control and the theories of the interactions of populations. *Parasitology* 31: 299–388.

Tiffney, B. H., and K. J. Niklas. 1985. Clonal growth in land plants: A paleobotan-ical perspective. pp. 35–66. In J. B. C. Jackson, L. W. Buss, and R. E. Cook (eds.). *Population biology and evolution of clonal organisms*. Yale University Press, New Haven.

Tilman, D. 1989. Discussion: Population dynamics and species interactions. pp. 89–100. In J. Roughgarden, R. M. May, and S. A. Levin (eds.). *Perspectives in ecological theory*. Princeton University Press, Princeton.

Tokuda, M., N. Maryana, and J. Yukawa. 2001. Leaf-rolling site preference by *Cycnotrachelus roelofsi* (Coleoptera: Attelabidae). *Entomol. Sci.* 4: 229–237.

Townes, H. 1969. *The genera of Ichneumonidae. Part I*. American Entomological Institute, Ann Arbor.

Tripp, H. A. 1957. Studies on the general biology and natural control of the jack pine sawfly, *Neodiprion swainei* Midd. *Can. Dept. Agr. For. Biol. Div., For. Biol. Lab. Quebec P.Q. Annu. Tech. Rep.* 1956: 1–36.

Uvarov, B. P. 1931. Insects and climate. *Trans. R. Entomol. Soc. London* 79: 1–247.

Uvarov, B. P. 1966. *Grasshoppers and locusts: A handbook of general acridology*, vol. 1. Cambridge University Press, Cambridge, U.K.

Uvarov, B. P. 1977. *Grasshoppers and locusts: A handbook of general acridology*, vol. 2. Centre for Overseas Pest Research, London.

Varley, G. C. 1949. Population changes in German forest insects. *J. Anim. Ecol.* 18: 117–122.

Varley, G. C. 1967. Estimation of secondary production in species with an annual life cycle. pp. 447–457. In K. Petrusewicz (ed.). *Secondary productivity in terrestrial ecosystems*. Institute of Ecology, Polish Academy of Sciences, Warsaw.

Varley, G. C. 1971. The effects of natural predators and parasites on winter moth populations in England. pp. 103–116. In *Proc. 2nd Tall Timbers Conf. on Ecol. Anim. Control by Habitat Management*.

Varley, G. C., and G. R. Gradwell. 1960. Key factors in population studies. *J. Anim. Ecol.* 29: 399–401.

Varley, G. C., and G. R. Gradwell. 1968. Population models for the winter moth. pp. 132–142. In T. R. E. Southwood (ed.). *Insect abundance*. Symp. R. Entomol. Soc. London 4. Royal Entomological Society, London.

Varley, G. C., and G. R. Gradwell. 1970. Recent advances in insect population dynamics. *Annu. Rev. Entomol.* 15: 1–24.

Varley, G. C., G. R. Gradwell, and M. P. Hassell. 1973. *Insect population ecology: An analytical approach*. Blackwell Scientific Publications, Oxford, U.K.

Verhulst, P. F. 1838. Notice sur la loi que la population suit dans son accroissement. *Correspond. Math. Phys.* 10: 113–121.

Vieira, E. M., I. Andrade, and P. W. Price. 1996. Fire effects on a *Palicourea rigida* (Rubiaceae) gall midge: A test of the plant vigor hypothesis. *Biotropica* 28: 210–217.

Wagner, M. R., S. K. N. Atuahene, and J. R. Cobbinah. 1991. *Forest entomology in west tropical Africa: Forest insects of Ghana*. Kluwer, Dordrecht, The Netherlands.

Walker, M., and T. H. Jones. 2001. Relative roles of top-down and bottom-up forces in terrestrial tritrophic plant–insect herbivore–natural enemy systems. *Oikos* 93: 177–187.

Waloff, Z. 1946. Seasonal breeding and migrations of the desert locust (*Schistocerca gregaria* F.) in eastern Africa. *Anti-Locust Mem.* 1: 1–74.

Wallner, W. E. 1987. Factors affecting insect population dynamics: Differences between outbreak and non-outbreak species. *Annu. Rev. Entomol.* 32: 317–340.

Waring, G. L., and N. S. Cobb. 1992. The impact of plant stress on herbivore population dynamics. pp. 167–226. In E. Bernays (ed.). *Insect-plant interactions*, vol. 4. CRC Press, Boca Raton.

Waring, G. L., and P. W. Price. 1988. Consequences of host plant chemical and physical variability to an associated herbivore. *Ecol. Res.* 3: 205–216.

Washburn, J. O., and H. V. Cornell. 1981. Parasitoids, patches, and phenology: Their possible role in the local extinction of a cynipid gall wasp population. *Ecology* 62: 1597–1607.

Watkinson, A. R. 1986. Plant population dynamics. pp. 137–184. In M. J. Crawley (ed.). *Plant ecology*. Blackwell Scientific Publications, Oxford, U.K.

Watson, A., and R. Moss. 1970. Dominance, spacing behaviour and aggression in relation to population limitation in vertebrates. pp. 167–220. In A. Watson (ed.). *Animal populations in relation to their food resources*. Blackwell Scientific Publications, Oxford, U.K.

Watt, A. S. 1947. Pattern and process in the plant community. *J. Ecol.* 35: 1–22.

Watt, A. D., and B. J. Hicks. 2000. A reappraisal of the population dynamics of the pine beauty moth, *Panolis flammea*, on lodgepole pine, *Pinus contorta*, in Scotland. *Popul. Ecol.* 42: 225–230.

Watt, A. D., S. R. Leather, M. D. Hunter, and N. A. C. Kidd (eds.). 1990. *Population dynamics of forest insects*. Intercept, Andover, U.K.

Weis, A. E., and A. Kapelinski. 1984. Manipulation of host plant development by a gall-midge *Rhabdophaga strobiloides*. *Ecol. Entomol.* 9: 457–465.

White, T. C. R. 1969. An index to measure weather-induced stress of trees associated with outbreaks of psyllids in Australia. *Ecology* 50: 905–909.

White, T. C. R. 1974. A hypothesis to explain outbreaks of looper caterpillars, with special reference to populations of *Selidosema suavis* in a plantation of *Pinus radiata* in New Zealand. *Oecologia* 16: 279–301.

White, T. C. R. 1984. The abundance of invertebrate herbivores in relation to the availability of nitrogen in stressed food plants. *Oecologia* 63: 90–105.

White, T. C. R. 1993. *The inadequate environment: Nitrogen and the abundance of animals*. Springer-Verlag, Berlin.

Whitham, T. G. 1978. Habitat selection by *Pemphigus* aphids in response to resource limitation and competition. *Ecology* 59: 1164–1176.

Whitham, T. G. 1980. The theory of habitat selection: Examined and extended using *Pemphigus* aphids. *Am. Nat.* 115: 449–466.

Whitney, H. S. 1982. Relationships between bark beetles and symbiotic organisms. pp. 183–211. In J. B. Mitton and K. B. Sturgeon (eds.). *Bark beetles in North American conifers: A system for the study of evolutionary biology.* University of Texas Press, Austin.

Whittaker, J. B., and N. P. Tribe. 1998. Predicting numbers of an insect (*Neophilaenus lineatus*: Homoptera) in a changing climate. *J. Anim. Ecol.* 67: 987–991.

Whittaker, R. H., S. A. Levin, and R. B. Root. 1973. Niche, habitat and ecotope. *Am. Nat.* 107: 321–338.

Wiegmann, B. M., C. Mitter, and B. Farrell. 1993. Diversification of carnivorous parasitic insects: Extraordinary radiation or specialized dead end? *Am. Nat.* 142: 737–754.

Williams, D. J. 1991. Superfamily Coccoidea. pp. 457–464. In I. D. Naumann (ed.). *The insects of Australia: A textbook for students and research workers.* Cornell University Press, Ithaca.

Williams, D. W., and A. M. Liebhold. 1995. Influence of weather on the synchrony of gypsy moth (Lepidoptera: Lymantriidae) outbreaks in New England. *Environ. Entomol.* 24: 987–995.

Williams, D. W., and A. M. Liebhold. 2000. Spatial synchrony of spruce budworm outbreaks in eastern North America. *Ecology* 81: 2753–2766.

Wolff, J. O. 1997. Population regulation in mammals: An evolutionary perspective. *J. Anim. Ecol.* 66: 1–13.

Woodman, R. L. 1990. Enemy impact and herbivore community structure: Tests using parasitoid assemblages, predatory ants, and galling sawflies on arroyo willow. PhD thesis, Northern Arizona University, Flagstaff.

Woodman, R. L., and P. W. Price. 1992. Differential larval predation by ants can influence willow sawfly community structure. *Ecology* 73: 1028–1037.

Woods, J. O., T. G. Carr, P. W. Price, L. E. Stevens, and N. S. Cobb. 1996. Growth of coyote willow and the attack and survival of a mid-rib galling sawfly, *Euura* sp. *Oecologia* 108: 714–722.

Wool, D. In press. Long-term temporal patterns of gall abundance of *Baizongia pistaciae* (Homoptera: Aphidoidea): Do temperature and rainfall play a role?

Wynne-Edwards, V. C. 1962. *Animal dispersion in relation to social behaviour.* Oliver and Boyd, Edinburgh, U.K.

Wynne-Edwards, V. C. 1964. Population control in animals. *Sci. Am.* 211(2): 68–74.

Wynne-Edwards, V. C. 1965. Self-regulating systems in populations of animals. *Science* 147: 1543–1548.

Yamazaki, K. 2001. Preference-performance linkage in the willow twig-galling agromyzid fly, *Hexomyza simplicoides* (Diptera: Agromyzidae) on the willow *Salix chaenomeloides*. *Entomol. Sci.* 4: 301–306.

Ylioja, T., H. Roininen, M. P. Ayres, M. Rousi, and P. W. Price. 1999. Host-driven population dynamics in an herbivorous insect. *Proc. Natl Acad. Sci. U.S.A.* 96: 10735–10740.

Yukawa, J., and H. Masuda. 1996. *Insect and mite galls of Japan in colors.* Zenkoku-Nouson-Kyouiku-Kyoukai, Tokyo. (in Japanese)

Zinsser, H. 1935. *Rats, lice and history.* Little, Brown, Boston.

Zwölfer, H., and W. Völkl. 1997. Einfluss der Verhaltens adulter Insekten auf Ressourcen-Nutzung und Populationsdynamik: Ein Drei-Komponenten-Modell der Populationsdichte-Steuerung. *Entomol. Gener.* 21(3): 129–144.

Author index

Taxonomic index

Abies spp. 225
Abies balsamea 146, 148, 224
Abies concolor 136, 148, 190
Acacia polyphylla 136
Acer pseudoplatanus 222
Acleris sp. 159
Acleris variana 225
Acrididae 17, 81, 163–6, 175, 176, 180, 181, 238
Acrocercops brongniadella 137
Adelges cooleyi 128, 130
Adelgidae 49, 128, 130
Aeneolaemia 171
Agromyza sp. 141
Agromyza frontella 139
Agromyzidae 128, 223
Aleurodidae 223
Aleurotrachelus jelenekii 223
Aleyrodidae 136
Alnus spp. 226
Alnus incana 138, 189
Alnus rubra 226
Alsophila pometaria 157
Altacinae 172
Amaryllidaceae 202
Amauronematus distinguensis 102
Amauronematus eiteli 102, 108, 112, 113, 115, 117
Amblypalpis olivierella 128, 131
Amorpha fruticosa 128
Amorpha shoot galler 128
amphibians 195, 213, 214
Anacamptodes clivinaria 157
Anacardium occidentale 33
Anacridium melanorhodon 165
Anadiplosis nr. venusta 127
Andraca bipunctata 227, 229
Andricus sp. 127
Andricus kollari 127
Andricus lignicola 127
Aneugmenus spp. 221, 224

angiosperm trees 188, 189, 190, 191, 192, 193, 225–8
Anguria 38
Anoplonyx sp. 190, 191
Anthocaris cardaminis 228
Anthomyiidae 223
Anthonomous grandis 12
ants 81, 160, 203
Aphididae 43, 49, 128, 129, 130, 136, 141, 166–70, 179, 180, 181, 222, 229, 238
Aphrophora pectoralis 136, 141, 171
Aphrophoridae 171–2
Arachnida 45
Archips sp. 159
Arctiidae 157, 175, 227
Ardis bruniventris 102, 123
armyworms 175, 179
arroyo willow see Salix lasiolepis
Aspilia foliacea 137, 139
Aulacizes sp. 137
Austriocetes cruciata 165
awafuki see Aphrophora pectoralis

bagworms 159
Baizongia pistaciae 221, 222
balsam fir 146, 148, 224
bark beetles 172–4, 223, 238
bats 195, 202
Bauhinia sp. 127
bee flies 202
bees 202
beet armyworm 175
beetles 12, 137, 141, 172–4, 203, 238
beets 33
Bemisia argentifolii 136
Betula sp. 190, 193
Betula pendula 139, 223
Betula pubescens 136, 137, 139, 189, 223, 226
Betula verucosa 189

Subject index

Printed in the United States
By Bookmasters